Springer Optimization and Its Applications

Volume 219

Aims and Scope

Optimization has continued to expand in all directions at an astonishing rate. New algorithmic and theoretical techniques are continually developing and the diffusion into other disciplines is proceeding at a rapid pace, with a spot light on machine learning, artificial intelligence, and quantum computing. Our knowledge of all aspects of the field has grown even more profound. At the same time, one of the most striking trends in optimization is the constantly increasing emphasis on the interdisciplinary nature of the field. Optimization has been a basic tool in areas not limited to applied mathematics, engineering, medicine, economics, computer science, operations research, and other sciences.

The series **Springer Optimization and Its Applications (SOIA)** aims to publish state-of-the-art expository works (monographs, contributed volumes, textbooks, handbooks) that focus on theory, methods, and applications of optimization. Topics covered include, but are not limited to, nonlinear optimization, combinatorial optimization, continuous optimization, stochastic optimization, Bayesian optimization, optimal control, discrete optimization, multi-objective optimization, and more. New to the series portfolio include Works at the intersection of optimization and machine learning, artificial intelligence, and quantum computing.

Volumes from this series are indexed by Web of Science, zbMATH, Mathematical Reviews, and SCOPUS.

Carlos A. S. Oliveira • Miltiades P. Pardalos
Editors

Handbook of Artificial Intelligence and Data Sciences for Routing Problems

Springer

Editors

Carlos A. S. Oliveira
Principal Inventive Scientist, Network Chief Technology Organization
AT&T Labs Inc.
Bellevue, WA, USA

Miltiades P. Pardalos
Dept. Industrial and Systems Eng.
Texas A&M University
College Station, TX, USA

ISSN 1931-6828 ISSN 1931-6836 (electronic)
Springer Optimization and Its Applications
ISBN 978-3-031-78261-9 ISBN 978-3-031-78262-6 (eBook)
https://doi.org/10.1007/978-3-031-78262-6

Mathematics Subject Classification: 90B18, 68T20, 90B06, 65K05, 68R01, 68Txx

This Springer imprint is published by the registered company Springer Nature Switzerland AG
The registered company address is: Gewerbestrasse 11, 6330 Cham, Switzerland

Preface

The field of artificial intelligence and optimization has rapidly evolved, becoming a cornerstone of modern science and engineering. The chapters in this book present an in-depth exploration of the intersection between machine learning, combinatorial optimization, and real-world applications, particularly in transportation and network design. As someone deeply involved in these domains, I find this compendium of knowledge both timely and essential.

From the foundational concepts of machine learning models to the advanced heuristic methods and reinforcement learning strategies, the book provides a comprehensive roadmap for solving complex problems. The discussion of Bayesian optimization, in particular, highlights the power of combining probabilistic models with efficient search strategies to optimize performance, a technique that has revolutionized various industries.

The case studies included are not mere academic exercises; they showcase the practical implementation of algorithms in predicting route sequences, traffic management, and even eco-friendly transportation. These examples illustrate the potential of AI to address some of the most pressing challenges of our time, from improving public transit systems to optimizing fleet management.

Topics

One of the key strengths of this book lies in its coverage of classical optimization problems, such as the Traveling Salesman Problem and Knapsack Problem, alongside cutting-edge techniques like Generative Adversarial Networks (GANs) and Variational Autoencoders. This juxtaposition offers readers a unique perspective on how traditional and modern methods can be integrated to solve problems more effectively.

The chapters on specific problems delve into common application across a range of areas, from job scheduling to network design. The detailed explanation of algorithms such as construction, local search, and post-processing demonstrates the versatility of metaheuristics in tackling diverse optimization challenges.

The chapters dedicated to AI techniques for transportation are particularly noteworthy. With the advent of autonomous vehicles and smart cities, the integration of AI in traffic management and public transit optimization is no longer a futuristic concept but a present-day reality. The book's exploration of safety, accident prevention, and eco-friendly transportation further emphasizes the critical role of AI in shaping the future of urban mobility.

Finally, the coverage of the Assignment Problems and their relationship to graph theory, heuristics, and metaheuristics provides a solid foundation for understanding the complexity of modern telecommunication networks. The detailed analysis of various algorithms, including Tabu Search, Genetic Algorithms, and GRASP, offers valuable insights into how these methods can be applied to solve real-world problems.

Book Audience

This book is dedicated to researchers, practitioners, and students who want to explore the intersection of AI, optimization, and practical applications. The breadth and depth of the topics covered ensure that readers will come away with an increased understanding of both the theoretical and practical aspects of these fields.

In this work, our goal is to provide mathematical and computational tools to handle AI and Optimization strategies for Routing Problems. This type of problem occurs in several areas, including transportation, logistics, computer networks, and industrial automation.

The book incorporates practical application in several areas. Problems as distinct as multicast routing, frequency assignment, and other problems occurring in real-world applications can be modeled and solved using AI techniques. As such, the present book can benefit several groups of readers that might be interested in the solution of routing problems as well as learning about applied research in modern AI algorithms.

First, researchers in optimization will find that this book provides a reference for standard approaches to the mathematical aspects of routing problems. We discuss not only traditional methods such as mixed integer programming but also AI approaches including heuristics, metaheuristics, genetic algorithms, and deep learning, that have proved successful in providing optimal or near optimal solutions to these problems.

For students, this book can be used to provide a comprehensive overview of the current research on optimization issues for AI algorithms for optimization. In each chapter we cover a particular aspect that can be further explored using the detailed list of references at the end of each chapter.

Finally, professionals working in areas such as telecommunications, transportation, logistics, circuit design, etc., will find a source of valuable information on algorithms and mathematical models for AI-based optimization. The book can be used as a starting point to explore existing implementations for common problems occurring in the industry.

Wellington, FL, USA
College Station, TX, USA
August, 2024

Carlos A. S. Oliveira
Miltiades P. Pardalos

Contents

Contributors

A. R. Pitombeira-Neto Department of Industrial Engineering, Federal University of Ceará, Fortaleza, Brazil

A. F. S. Veloso Department of Computing, Federal University of Piauí (UFPI), Teresina, PI, Brazil

J. V. R. Júnior Department of Computing, Federal University of Piauí (UFPI), Teresina, PI, Brazil

M. M. do N. Costa Department of Software Engineering, iCEV Faculty (iCEV), Teresina, PI, Brazil

R. A. L. Rabelo Department of Computing, Federal University of Piauí (UFPI), Teresina, PI, Brazil

P. R. Pinheiro Department of Computer Science, University of Fortaleza (UNIFOR), Fortaleza, CE, Brazil

R. S. Iwayama Graduate Program in Logistics Systems Engineering, University of São Paulo, São Paulo, SP, Brazil

C. B. Cunha Graduate Program in Logistics Systems Engineering, University of São Paulo, São Paulo, SP, Brazil

Universidade de São Paulo, Departamento de Eng de Transportes - Escola Politécnica, Ed Paula Souza (Prédio da Eng Civil) Cidade Universitária, São Paulo, Brazil

M. Negreiros Universidade Estadual do Ceará, Laboratório de Computação Científica, Av. Silas Munguba, 1700 (Campus do Itaperi), Fortaleza, Brazil

N. Maculan Universidade Federal do Rio de Janeiro, Programa de Eng de Sistemas e Computação (PESC-COPPE), Rio de Janeiro, RJ, Brazil

C. A. S. Oliveira Principal Inventive Scientist, Network Chief Technology Organization, AT&T Labs Inc., Bellevue, WA, USA

F. H. Viana Centro Federal de Educação Tecnológica Celso Suckow da Fonseca CEFET/RJ, Campus de Nova Iguaçu Laboratório de Mobilidade Urbana Rio de Janeiro, Rio de Janeiro, Brazil

F. L. Mello Universidade Federal do Rio de Janeiro, Av. Brigadeiro Trompowiski s/n, Centro de Tecnologia, Bloco H, sala H-212b Cidade Universitária, Ilha do Fundão, Rio de Janeiro, RJ, Brazil

Acronyms

ACO	Ant Colony Optimization
ADAS	Advanced Driver Assistance Systems
AMPS	Advanced Mobile Phone System
AUC	Area Under the Curve
BKS	Best-Known Solution
BPCC	Bin Packing Problem with Compatible Categories
BPP	Bin Packing Problems
CCP	Capacitated Clustering Problem
CDMA	Code Division Multiple Access
DCMRP	Delay-Constrained Multicast Routing Problem
DFAP	Decision Frequency Assignment Problem
FA	Frequency Assignment
FF	First Fit
GAN	Generative Adversarial Networks
GGA-CGT	Grouping Genetic Algorithm with Controlled Gene Transmission
GRASP	Greedy Randomized Adaptive Search Procedure
GSM	Global Systems for Mobile Communication
ILS	Iterated Local Search
IMC	Intermodal Carrier
IRL	Inverse Reinforcement Learning
MAVRP	Multi-Attribute Vehicle Routing Problems
MILP	Mixed-Integer Linear Programming
MISP	Maximum Independent Set Problem
MNL	Multinomial Logit Model
MRP	Multicast Routing Problem
MSPT	Multi Start Perturbation Tabu Neighborhood Search
PGOCCDD	Problem of Grouping Orders into Shipments with Compatible Categories and Freight Based on the Direct Distances to the Farthest Customers
PSO	Particle Swarm Optimization
RLC	Restricted List of Candidates

SI	Swarm Intelligence
SMDD	Sum of the Maximum Direct Distance
TDMA	Time Division Multiple Access
TMS	Transportation Management Systems
TSPTW	Transportation Management Systems with Time Windows
TSP	Traveling Salesman Problem
TS	Tabu Search
VAE	Variational Autoencoders
VRP	Vehicle Routing Problems

List of Figures

List of Tables

Chapter 1
Route Sequence Prediction Through Inverse Reinforcement Learning and Bayesian Optimization

Anselmo R. Pitombeira-Neto

Abstract We consider the problem of predicting the route sequence a driver will follow given a set of stops and additional data on the delivered packages. We propose an innovative approach to route sequence prediction by using inverse reinforcement learning and Bayesian optimization. Given a sample of route sequences followed by a set of drivers, we want to learn a parameterized cost function that drivers are trying to minimize. Predicted route sequences are generated by using a rollout algorithm. We define our loss function as the average score over predicted route sequences from training data. As the loss function is non-differentiable, we resort to Bayesian optimization, which allows us to efficiently find good parameter values given a limited computational budget. We illustrate the proposed approach with the use of real data and show that it can reasonably approximate real routes generated by human drivers. This suggests that it may be possible to develop models which can learn local information available to drivers and which is reflected in the real routes they build. In some applications, especially in urban environments, routes planned by a machine learning model may be more feasible than routes planned by an optimization model.

Keywords Routing · Reinforcement learning · Bayesian optimization

1.1 Introduction

Companies which deliver products to customers often face the problem of planning the routes that delivery vehicles have to follow in order to meet demand. Given data on the stops to be visited, demanded quantities, and capacity of vehicles, among other characteristics and constraints, it is possible to formulate a mathematical programming model whose solution optimizes some objective function, such as

---

A. R. Pitombeira-Neto (✉)
Department of Industrial Engineering, Federal University of Ceará, Fortaleza, Brazil
e-mail: anselmo.pitombeira@ufc.br

C. A. S. Oliveira, M. P. Pardalos (eds.), *Handbook of Artificial Intelligence and Data Sciences for Routing Problems*, Springer Optimization and Its Applications 219, https://doi.org/10.1007/978-3-031-78262-6_1

minimization of total length of routes, travel time, or cost. There is a large body of literature dedicated to analyzing and proposing solution methods to the various *vehicle routing problems* (VRP), according to characteristics of different particular applications.

Although VRP models have been successful in many real applications, it is hard to consider in these models all the range of local information which is available to drivers, such as traffic conditions, availability parking lots, accessibility of urban areas, and weather conditions. As a consequence, in practice drivers often modify the routes suggested by optimization software or plan their own routes from scratch in order to take into account available local information.

1.1.1 Machine Learning Models

Unlike the classical VRPs, in which one is interested in designing optimal routes, we are interested in developing a machine learning or statistical model which may be capable of predicting routes similar to the ones generated by human drivers. Such a model may be trained from a sample of real observed routes and *learn* to generate routes similar to the ones created by human drivers. The advantage of these routes over the ones suggested by optimization models is that they may be more feasible in practice, since they take into account the local information known by drivers which is hard to consider in optimization models.

In particular, we tackle the problem of predicting the route sequence a driver will follow given a set of required stops and additional data on the delivered packages. More precisely, let $\mathcal{S}$ be a finite set of stops corresponding to locations a driver has to visit to deliver packages. The driver will choose a sequence of stops

$$r = (s_1, s_2, \ldots, s_n, s_{n+1}),$$

in which $s_t, t = 1, \ldots, n+1$ is a stop in the sequence, s_1 is the starting location (often called the depot), and $s_{n+1} = s_1$ (the driver finishes the route in the depot). Given the set of stops $\mathcal{S}$ and data $\mathcal{D}$, we want to predict the route sequence r. $\mathcal{D}$ may include data on the stops, such as the geographical coordinates of the stops, their types, and zip codes, as well as data on the packages, such as size, weight, and time windows, among others. The set $\mathcal{S}$ and data $\mathcal{D}$ are unique to each route. This route sequence prediction problem has been largely unexplored in the literature.[1]

[1] The route sequence prediction problem was motivated by a real case from Amazon and addressed in the Amazon Last Mile Routing Research Challenge. See https://routingchallenge.mit.edu/.

1.1.2 Proposed Algorithm

We propose an innovative approach to route sequence prediction by using inverse reinforcement learning (IRL) and Bayesian optimization. Reinforcement learning is a technique related to dynamic programming in which we want to find a possibly optimal decision policy for an intelligent agent such that it accomplishes some desired task [21]. The desired behavior is achieved through the design of a reward (or cost) function. In contrast, in IRL we are given a sample of demonstrations from one or more intelligent agents and we want to learn a reward function compatible with the behavior of the agents [1, 17].

From the point of view of a single driver, the problem of generating a route sequence is very similar to a traveling salesman problem (TSP) with time windows (TSPTW). It is well known from the literature that the TSP may be formulated as a dynamic program, in which the states are partial sequences of stops and the cost function is assumed to be the distance (or time) between a current stop and the possible next stops [5, 15]. In our case, we want to solve the inverse problem: we have demonstrations of solutions to the TSPTW generated by a set of drivers, and we want to learn an approximation of an unknown cost function that the drivers are trying to minimize. If we can reasonably approximate this cost function, we may try to use dynamic programming to generate a route sequence given a set of stops, which could then be used as the predicted sequence of stops a driver would follow.

1.1.3 Linear Model

In order to approximate the unknown cost function, we use a linear model and try to learn the parameters from the sample of route sequences. However, we face some hurdles: in theory, we may use maximum likelihood (or even Bayesian inference) to learn the parameters, but trying to compute a likelihood function demands specifying a probability measure over all possible TSPTW solutions, which is computationally intractable. We then define our loss function as the average score over route sequences in a training sample and try to set the parameters so as to minimize the average score. The score function measures the similarity between observed and predicted route sequences.

Moreover, in order to compute the score, we need to generate route sequences by solving the TSPTW with given parameters for the cost function. However, large TSPs cannot be solved by exact dynamic programming due to the exorbitant state space. We solve them heuristically by using a *rollout* algorithm, which is a kind of approximate dynamic programming. Finally, as the loss function is non-differentiable, this precludes us from using gradient-based methods. (And computing finite differences will be overkill due to the large time to evaluate the score function in the whole sample.) We then resort to Bayesian optimization, which allows us to efficiently find good parameter values given a limited computational budget.

In Sect. 1.2, we provide some background which supports our approach to the route sequence prediction problem; in Sect. 1.3 we detail the methodology; in Sect. 1.4 we describe the application of our approach to real data on observed route sequences; finally, in Sect. 1.5 we draw some conclusions and give some recommendations on further research.

1.2 Related Work

Since the route sequence prediction problem has been largely unexplored in the literature, we refer below to related works which may provide some background and context to our approach to the problem.

VRPs and TSPs (many of which are particular cases of the more general VRP class of problems) are among the most studied problems in operations research. Its most basic version involves the planning of a set of finite routes, each corresponding to a subset of customers to be visited, subject to vehicles' capacity constraints. The objective is often to plan routes with least total distance or time.

1.2.1 Heuristic Methods

The problem is solved with the use of mathematical programming or heuristic algorithms. There is a large body of literature regarding many different variants of VRPs and TSPs; we refer to the works by Applegate et al. [2] and Toth and Vigo [22] for excellent surveys on the subject. As emphasized in the introduction of this chapter, the main limitation of the routes planned by operations research models is that they do not take into account local and contingent information which is available to drivers.

Dynamic programming (DP) is a well-known technique for solving sequential decision problems proposed by Bellman [4]. In contrast to mathematical programming, which produces optimal *open loop* controls for a decision problem, the objective in DP is to find a decision policy, i.e., *closed loop* controls. A decision policy specifies which decision to make at each possible state of a system or environment. DP suffers from the *curse of dimensionality*, which refers to the combinatorial explosion of the number of states of a system as the dimension of the state vector increases. In addition, DP assumes that a model of the system is available, which often is not the case.

1.2.2 Reinforcement Learning

Reinforcement learning (RL) is a machine learning technique whose mathematical framework is essentially the same as DP, but which tries to overcome the abovementioned limitations of DP by using approximation techniques such as Monte Carlo simulation, artificial neural networks, etc. [21]. Many of these techniques have been concurrently developed by operational researchers under the name *approximate dynamic programming* [13, 20].

A noteworthy technique for approximately solving DP/RL problems is called the *rollout* algorithm. In DP/RL, an optimal policy is obtained if one knows the optimal value function, which may be computed by backward dynamic programming in the case of finite-horizon problems. The rollout algorithm is a kind of *forward* dynamic programming, in which the value function of a state in the next decision epoch is computed heuristically on the fly [7, 8]. It has been applied to both the TSP [14] and the TSPTW [11]. It is a component of the Monte Carlo tree search technique, which has been successfully applied to games [10].

1.2.3 Reward Function

In RL/DP, the reward/cost function is frequently specified to obtain a desired solution of the sequential decision problem. In contrast, in *inverse reinforcement learning* (IRL), we are interested in estimating a reward/cost function from a sample of observed trajectories produced by intelligent agents [1, 17]. IRL has been recently applied to medical decisions [6] and to trajectory modeling [19]. A recent survey on IRL is presented by Arora and Doshi [3].

Our approach draws from ongoing work from the author on modeling sparse trajectories with the use of IRL [19]. But, in contrast to this latter work, in which value functions may be computed exactly, the state space in the route sequence prediction problem is too large. The works on the application of the rollout algorithm to the TSP and the TSPTW sparked the idea of using a parameterized rollout algorithm to approximate the value function. As far as we know, our approach has not been tried before in the route sequence prediction problem.

1.3 Methodology

The core idea of our approach is to iteratively alternate between a route sequence prediction step and a parameter update step. For a given data set related to a sample of observed route sequences, we apply a parameterized route sequence prediction algorithm and compute the score of each predicted route sequence. We then apply a Bayesian optimization step to suggest a new set of parameter values with the objective of minimizing the average score over the sample. The two steps are iterated until a predefined number of iterations is reached.

1.3.1 Prediction of a Route Sequence Given Data

Given a set of stops $\mathcal{S}$ and related data $\mathcal{D}$, we assume a driver is an optimizer agent and tries to find a route sequence $r = (s_1, s_2, \ldots, s_n, s_1)$ so as to minimize some cost function unknown to an external observer, in which the starting s_1 is fixed and called the *station* and $n = |\mathcal{S}|$. We define as the state $x_t \in \mathcal{X}_t$ at a decision epoch t a partial sequence:

$$x_t := (s_1, s_2, \ldots, s_t), \quad 1 \leq t < n, \tag{1.1}$$

in which $\mathcal{X}_t$ is the set of all combinations of stop sequences of size t. At a state x_t, a driver chooses a next stop a_t from a set of remaining stops $\mathcal{A}(x_t) = \mathcal{S} \setminus \{x_t\}$ and then incurs an immediate cost given by

$$c(x_t, a_t) = \sum_{k=1}^{m} \phi_k(x_t, a_t)\beta_k, \tag{1.2}$$

in which $\phi_k(x_t, a_t), k = 1, \ldots, m$ represent features associated with the current partial sequence and the next stop, and $\beta_k, k = 1, \ldots, m$ are parameters. We assume that features may also be functions of the data $\mathcal{D}$ related to the stops.

1.3.2 Solving the Model

Notice that if we know the values of parameters $\beta_k, k = 1, \ldots, m$, we may (in theory) find an optimal route sequence by solving the following optimal control problem:

$$\min_{a_1, a_2, \ldots, a_{n-1}} \sum_{t=1}^{n-1} \alpha^t c(x_t, a_t) + \alpha^n c(x_n, s_1), \tag{1.3}$$

$$\text{s.t.} \tag{1.4}$$

$$x_{t+1} = (x_t, a_t), \quad t = 1, \ldots, n-1, \tag{1.5}$$

$$a_t \in \mathcal{A}(x_t), \quad t = 1, \ldots, n-1, \tag{1.6}$$

$$x_t \in \mathcal{X}_t, \quad t = 2, \ldots, n-1, \tag{1.7}$$

$$x_1 = s_1, \tag{1.8}$$

in which $0 < \alpha \leq 1$ is a discount factor which controls "how far into the future" the agent plans. The optimization problem (1.3) may be solved by backward dynamic programming by recursively solving Bellman's equation

$$v_t(x_t) = \min_{a_t \in \mathcal{A}(x_t)} \{c(x_t, a_t) + \alpha v_{t+1}(x_{t+1})\}, \quad \forall x_t \in \mathcal{X}_t, \quad t = 1, \ldots, n-1, \tag{1.9}$$

in which $v_t(x_t)$ is the value function associated with state x_t (also called a *cost-to-go* function) and $v_n(x_n) = c(x_n, s_1)$. However, it will be computationally infeasible for a large set of stops, so we resort to a rollout algorithm, which is a form of forward dynamic programming.

In the rollout algorithm, at a given state x_t, a decision $a_t^\star$ is made according to

$$a_t^\star \in \arg\min_{a_t \in \mathcal{A}(x_t)} \{c(x_t, a_t) + \alpha h_{t+1}(x_{t+1})\}, \tag{1.10}$$

in which $h_{t+1}(x_{t+1})$ is an estimate of the value function $v_{t+1}(x_{t+1})$ obtained by generating a sequence of stops by following a heuristic decision policy. We will use a greedy policy in relation to the cost function $c(x_t, a_t)$ as a base heuristic, i.e., at each state x_t the greedy policy chooses the next stop which minimizes the immediate cost. The rollout algorithm is guaranteed to obtain solutions at least as good as the base policy (under some conditions) due to the policy improvement property of dynamic programming [9].

1.3.3 Tracking Time

In addition, due to the existence of time windows, we maintain a clock variable Γ which tracks the current time in the route sequence. We denote the left bound and the right bound of a time window at a stop $s \in \mathcal{S}$ as $w^-(s)$ and $w^+(s)$, respectively. The purpose of this clock variable is to check it against the time windows and use this information in the cost function. It is initialized to the departure time and updated by the service times and travel times along the route.

An important step is the definition of features to be used in the cost function (1.2). At a given state $x_t = (s_1, s_2, \ldots, s_t)$ and for each possible next stop $a_t \in \mathcal{A}(x_t)$, we have defined the following features according to the available data on the stops and packages:

1. $\phi_1(x_t, a_t)$—The travel time $\tau_1(x_t, a_t)$ between the current stop s_t and a possible next stop a_t
2. $\phi_1(x_t, a_t)$—The service time $\tau_2(a_t)$ in a next stop a_t
3. $\phi_3(x_t, a_t)$—The maximum dimension size of all packages in a next stop
4. $\phi_4(x_t, a_t)$—The time difference between the starting of the time window of a next stop and the current clock, truncated to zero if the difference is negative, i.e., $\max(0, w^-(a_t) - \Gamma)$
5. $\phi_5(x_t, a_t)$—The time difference between the end of the time window of a next stop and the current clock, i.e., $\Gamma - w^+(a_t)$

Algorithm 1 describes in detail the rollout algorithm to compute a route sequence, while Algorithm 2 describes the procedure to compute the approximate value

Algorithm 1 Rollout algorithm for computing a route sequence

Input: Stops $\mathcal{S}$, station s_1, data $\mathcal{D}$, features $\phi(.)$, parameters θ

$\mathcal{S}' \leftarrow \mathcal{S}$
$\Gamma \leftarrow$ departure time at s_1
$x \leftarrow [s_1]$
while $\mathcal{S}' \neq \emptyset$ **do**
  $q^\star \leftarrow +\infty$
  for $a \in \mathcal{S}'$ **do**
    Compute $c(x, a) = \sum_{k=1}^{m} \phi_k(x, a)\beta_k$
    $x' \leftarrow [x, a]$
    $\Gamma' \leftarrow \Gamma + \tau_1(x, a) + \tau_2(a)$ ▷ Tentative clock
    Compute approximate value function $h(x'; \Gamma')$ ▷ Run greedy heuristic in Algorithm (2)
    $q(x, a) \leftarrow c(x, a) + \alpha h(x')$
    if $q(x, a) < q^\star$ **then**
      $q^\star \leftarrow q(x, a)$
      $a^\star \leftarrow a$
    end if
  end for
  $\Gamma \leftarrow \Gamma + \tau_1(x, a^\star) + \tau_2(a^\star)$ ▷ Update route clock
  $x \leftarrow [x, a^\star]$
  $\mathcal{S}' \leftarrow \mathcal{S}' \setminus \{a^\star\}$
end while
return x ▷ generated route sequence

function $h(x)$ at a state x. Notice that in Algorithm 2 we penalize $h(x)$ by a penalty P in case a time window is not met and $\gamma \geq 0$ is a parameter which adjusts the magnitude of the penalty.

1.3.4 Learning the Parameters of the Prediction Function

We define a prediction function

$$\hat{r} = f(\mathcal{S}, \mathcal{D}, \theta) \tag{1.11}$$

which produces a predicted route sequence $\hat{r}$ given the set of stops and related data by applying the rollout algorithm given in Algorithm 1, in which $\theta = (\alpha, \beta_1, \beta_2, \ldots, \beta_m, \gamma)$ are parameters. We use this function to define the problem of predicting route sequences as a supervised learning task.

First, we need to define a scoring metric, which measures the quality of a prediction. Given an observed route sequence r generated by a driver and a corresponding predicted route sequence $\hat{r}$ given by the model, a score function is defined as

Algorithm 2 Greedy heuristic for approximating the value function at a state x

Input: Partial route x, stops $\mathcal{S}$, station s_1, current tentative clock time Γ', windows penalty P, features $\phi(.)$, parameters θ, data $\mathcal{D}$
$\mathcal{S}' \leftarrow \mathcal{S} \setminus \{x\}$
$x' \leftarrow x$
$h(x) \leftarrow 0$
while $\mathcal{S}' \neq \emptyset$ **do**
  $c^\star \leftarrow +\infty$
  for $a \in \mathcal{S}'$ **do**
    Compute $c(x', a) = \sum_{k=1}^{m} \phi_k(x', a)\beta_k$
    if $c(x', a) < c^\star$ **then**
      $c^\star \leftarrow c(x', a)$
      $a^\star \leftarrow a$
    end if
  end for
  $\Gamma' \leftarrow \Gamma' + \tau_1(x', a^\star) + \tau_2(a^\star)$ ▷ Update route clock
  $x' \leftarrow [x', a^\star]$
  $h(x) \leftarrow h(x) + c^\star$
  if $\Gamma' < w^-(a^\star)$ or $\Gamma' > w^+(a^\star)$ **then**
    $h(x) \leftarrow h(x) + \gamma P$ ▷ Penalty for violating time window
  end if
  $\mathcal{S}' \leftarrow \mathcal{S}' \setminus \{a^\star\}$
end while
$h(x) \leftarrow h(x) + c(x', s_1)$
return $h(x)$

$$\text{score}(\hat{r}, r) = \frac{\text{SD}(\hat{r}, r) \times \text{ERP}_{\text{norm}}(\hat{r}, r)}{\text{ERP}_{\text{e}}(\hat{r}, r)}. \tag{1.12}$$

SD measures the *sequence deviation* between predicted and observed route sequences. It is computed by first building a list b in the following manner: b is initially empty; then, go through sequence r and for each stop identify its corresponding position in sequence $\hat{r}$, appending its position to list b. After that, SD may be computed by

$$\text{SD} = \frac{2}{n(n-1)} \sum_{i=1}^{n} |b_i - b_{i-1}| - 1, \tag{1.13}$$

in which n is the number of stops in the route sequence. ERP_{norm} stands for *normalized edit distance with real penalty* (a penalized form of the Levenshtein distance) and ERP_{e} is the number of edit operations required to transform $\hat{r}$ in r when computing ERP_{norm}.

Given a set of observed route sequences $\{r_j\}_{j=1}^{J}$ and accompanying data $\mathcal{D}_j$, we define as our *loss function* the average score over all stop sequences:

$$\ell(\theta) = \frac{1}{J}\sum_{j=1}^{J} \mathrm{score}(\hat{r}_j, r_j), \tag{1.14}$$

in which $\hat{r}_j = f(\mathcal{S}_j, \mathcal{D}_j, \theta)$. Our problem is then to determine the parameter values $\theta^\star$ so that

$$\theta^\star \in \arg\min_{\theta\in\Theta} \ell(\theta), \tag{1.15}$$

in which Θ is an appropriate parameter space. As the loss function (1.14) is non-differentiable, we resort to a gradient-free method. Moreover, it is computationally expensive, so that using finite differences or random-search-based methods such as genetic algorithms will be overkill. We then apply Bayesian optimization, which is a technique designed for searching global optima of computationally expensive functions [12].

1.3.5 Bayesian Optimization

Bayesian optimization is a derivative-free technique for efficient optimization of expensive functions. Let $\ell(\theta)$ be the function that we want to find a minimum. We first specify a Gaussian process on the space of continuous functions $\ell : \Theta \to \mathbb{R}$ (notice that we do not know the exact shape of the function ℓ). The Gaussian process encodes our prior belief about the function ℓ and it is defined by a mean function $\mu_0(\theta)$ and a positive-definite kernel function $\Sigma_0(\theta, \theta')$. $\mu_0(\theta)$ specifies our initial belief on the expected value of $\ell(\theta)$ as θ varies, while $\Sigma_0(\theta, \theta')$ specifies our belief about the covariance between any two values $\ell(\theta)$ and $\ell(\theta')$. Kernel functions assume that the covariance between two values $\ell(\theta)$ and $\ell(\theta')$ is inversely proportional to the distance between θ and θ'. There are many different kernel functions proposed in the literature.

We start by sampling a set of n initial evaluation points $\{\theta_1, \theta_2, \ldots, \theta_n\}$ according to some procedure (e.g., sampling randomly or using some space-filling procedure). We will use by convenience the compact notation $\theta_{1:n} = (\theta_1, \theta_2, \ldots, \theta_n)$ and $\ell(\theta_{1:n}) = (\ell(\theta_1), \ell(\theta_2), \ldots, \ell(\theta_n))$. By substituting this finite set of values into the mean and kernel functions, we find a prior mean vector $\mu_0(\theta_{1:n})$ and a covariance matrix $\Sigma_0(\theta_{1:n}, \theta_{1:n})$, whose entries are given by the kernel function (notice there is some abuse of notation here, since we use the same symbols for both the kernel function and the covariance matrix). Thus, the Gaussian process reduces to a multivariate normal probability density function. We then evaluate our score function at each point and denote by $\mathcal{F} = \{(\theta_1, \ell_1), (\theta_2, \ell_2), \ldots, (\theta_n, \ell_n)\}$ the set of evaluation points and corresponding function values, in which $\ell_j = \ell(\theta_j)$. We then update our belief about $\ell(\theta)$ by obtaining the *posterior* distribution $p(\ell(\theta)|\mathcal{F})$.

From the properties of the multivariate normal distribution (see, e.g., Chapter 17 of [23]), we have that the posterior is a univariate normal distribution with

$$\mu_n(\theta) = \mu_0(\theta) + \Sigma_0(\theta, \theta_{1:n})\Sigma_0(\theta_{1:n}, \theta_{1:n})^{-1}(\ell(\theta_{1:n}) - \mu_0(\theta_{1:n})), \tag{1.16}$$

$$\sigma^2(\theta) = \Sigma_0(\theta_{1:n}, \theta_{1:n}) - \Sigma_0(\theta, \theta_{1:n})\Sigma_0(\theta_{1:n}, \theta_{1:n})^{-1}\Sigma_0(\theta_{1:n}, \theta). \tag{1.17}$$

The function $\mu_n(\theta)$ gives our expected posterior belief of where the function value $\ell(\theta)$ is located and $\sigma^2(\theta)$ gives our uncertainty.

1.3.6 Using Acquisition Function

We may now use the posterior distribution $p(\ell(\theta)|\mathcal{F})$ to find a minimum for $\ell(\theta)$, but since we are uncertain about the exact shape of it, Bayesian optimization uses an *acquisition function* $\hat{f}(p(\ell(\theta)|\mathcal{F}))$ to balance exploration and exploitation, i.e., it tries to balance the effort between sampling candidate points in regions of the domain with high uncertainty about the values of $\ell(\theta)$ and in regions which are reasonably known to provide good values of $\ell(\theta)$. A candidate point θ_{n+1} is obtained by optimizing the acquisition function. There are many acquisition functions proposed in the literature. For example, we may find as a candidate point the one which maximizes the *expected improvement*:

$$\theta_{n+1} = \arg\max_{\theta \in \Theta} \mathbb{E}[\max(\ell^\star - \ell(\theta), 0)], \tag{1.18}$$

in which $\ell^\star = \min_{j=1,\ldots,n}(\ell_j)$ is the best value found so far, $\max(\ell^\star - \ell(\theta), 0)$ is the improvement over $\ell^\star$, and the expected value is computed relative to the posterior distribution $p(\ell(\theta)|\mathcal{F})$. More details about the expected improvement acquisition function may be found in [16].

Bayesian optimization of our loss function may be then summarized by the following steps:

1. Set a prior Gaussian process $p(\ell(\theta))$ over the unknown function $\ell(\theta)$ by specifying a mean function $\mu_0(\theta)$ and kernel function $\Sigma_0(\theta, \theta')$.
2. Sample a set of n initial evaluation points and corresponding values $\mathcal{F} = \{(\theta_1, \ell_1), (\theta_2, \ell_2), \ldots, (\theta_n, \ell_n)\}$ according to some procedure (e.g., sampling randomly or using some space-filling procedure).
3. Obtain the posterior distribution $p(\ell(\theta)|\mathcal{F})$ by computing $\mu_n(\theta)$ and $\sigma_n^2(\theta)$ from Eqs. (1.16) and (1.17).
4. Obtain a new candidate point θ_{n+1} by maximizing the expected improvement (Eq. (1.18)).
5. Update $\mathcal{F} \leftarrow \mathcal{F} \cup \{(\theta_{n+1}, \ell(\theta_{n+1}))\}$.

Algorithm 3 Bayesian optimization of the loss function $\ell(\theta)$

Input: Stops $\mathcal{S}$, station s_1, data $\mathcal{D}$, features $\phi(.)$, I_0, I, J, $p(\ell)$
Compute I_0 random initial points $\mathcal{F} = \{(\theta_1, \ell_1), (\theta_2, \ell_2), \ldots, (\theta_{I_0}, \ell_{I_0})\}$
$\ell^\star \leftarrow +\infty$
for $i = I_0 + 1, \ldots, I$ **do**
  Update posterior distribution $p(\ell(\theta)|\mathcal{F})$ ▷ Eqs. (1.16) and (1.17)
  Generate a candidate θ_i by optimizing an acquisition function (e.g. expected improvement)
  for $j = 1, \ldots, J$ **do**
    Compute predicted route sequence $\hat{r}_{ij} = f(\mathcal{S}_j, \mathcal{D}_j, \theta_i)$ ▷ By using Algorithm 1
  end for
  Compute loss function $\ell_i = \frac{1}{J}\sum_{j=1}^{J} \text{score}(\hat{r}_{ij}, r_j)$ ▷ r_j is the observed route sequence
  if $\ell_i < \ell^\star$ **then**
    $\ell^\star \leftarrow \ell_i$
    $\theta^\star \leftarrow \theta_i$
  end if
  $\mathcal{F} \leftarrow \mathcal{F} \cup \{(\theta_i, \ell_i)\}$
end for
return $\theta^\star$

6. Repeat 3, 4, and 5 until a stopping criterion is met. Return $\theta^\star$ corresponding to the best value $\ell(\theta^\star) = \min\{\ell(\theta_1), \ell(\theta_2), \ldots, \ell(\theta_I)\}$ where I is the total number of iterations.

The Bayesian optimization algorithm used to minimize the loss function (1.14) is summarized in Algorithm 3.

1.4 Case Study

In this section, we illustrate the application of the proposed approach to data made available by the Amazon Last Mile Routing Research Challenge.[2] The data comprises information on approximately 6000 routes generated by Amazon drivers, including information on the particular stops and on the delivered packages.

Information on the routes includes a route id, the date in which the route was executed, its departure time, capacity of the vehicle, the set of stops to be served, the real observed sequence of stops, and a label which classifies the quality of a route in high, medium or low. Information on the stops includes a unique id, the geographical coordinates of the stop, the type of stop, its zone id, the set of packages to be delivered, and all distances to any other stop on the same route. Finally, information on packages includes a unique id, the status of the package (delivered or not), a time window, the planned service time, and the geometric dimensions of the package.

Regarding data treatment, we applied almost no preprocessing. Since we are interested in predicting high-quality route sequences, we filtered only the 2718

[2] https://routingchallenge.mit.edu/.

high-quality routes available in the data and split in approximately 70% training sample (1903 routes) and 30% testing sample (815 routes). We converted the departure times and time windows to POSIX time (in seconds), so that we could appropriately track the route clock. The data was not normalized. In preliminary experiments normalization did not exhibit better performance, although we do think more experiments are needed to evaluate if normalization has an effect.

1.4.1 *Results*

We applied Bayesian optimization (Algorithm 3) for a fixed number of 30 iterations to the training sample and retained the parameter values which achieved the lowest average score over the training sample. The model has seven parameters, five βs directly related to the features in the cost function (1.2), the discount factor α, and the penalty parameter γ. All parameters were assumed to be bounded in the [0, 1] interval. We have used the expected improvement acquisition function given in (1.18).

We have also tested the *upper confidence bound* and *maximum probability of improvement* acquisition functions, but they produced worse results in preliminary experiments. We used the `bayesian-optimization` Python library [18]. We started at a greedy solution with parameter values $\beta_1 = 1$ and all other parameters equal zero, whose average score is 0.1028. Each iteration took around 40 to 60 minutes in an Intel i5-9600K CPU with 16 GB RAM machine. Thus, the computation of all iterations took almost 30 clock hours.

Figure 1.1 shows the average score for all 30 iterations. The parameter values with lowest average score (= 0.09055) were found in iteration 26. As a benchmark, perfect prediction has a score with value 0.0, while a random predictor (just drawing the route sequence randomly) would show scores between 0.8 and 1.2. Obtained parameter values were $\beta_1 = 1.0$, $\beta_2 = 0.0$, $\beta_3 = 0.3047$, $\beta_4 = 0.0$, $\beta_5 = 0.0$, $\alpha = 0.377$, $\gamma = 0.0$. Notice that travel time (corresponding to parameter β_1) is the main explanatory feature, while features related to time windows were zeroed. We think this is an artifact of this particular solution, and suspect better solutions should attribute some weight to features related to time windows. Parameter α indicates that drivers seem to look some steps into the future and therefore are not myopic.

1.4.2 *Predicting Route Sequences*

We then applied Algorithm 1 to predict each of the 815 routes sequences in the testing sample with the parameters found during the training step. Application of Algorithm 1 is very fast and we could generate route sequence predictions for all 815 routes in less than 5 minutes. The average score obtained in the testing sample was 0.0958 and median score was 0.0823. Figure 1.2 shows the distribution of the

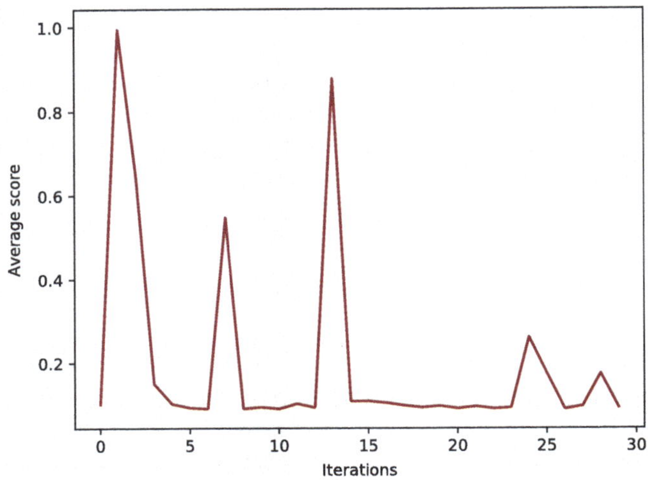

Fig. 1.1 Iterations from the Bayesian optimization algorithm over the training sample. Lowest average score (= 0.09055) was found in iteration 26. Peaks are due to the Bayesian optimization exploration of the parameter space

Fig. 1.2 Score distribution of 815 predicted route sequences in the holdout sample

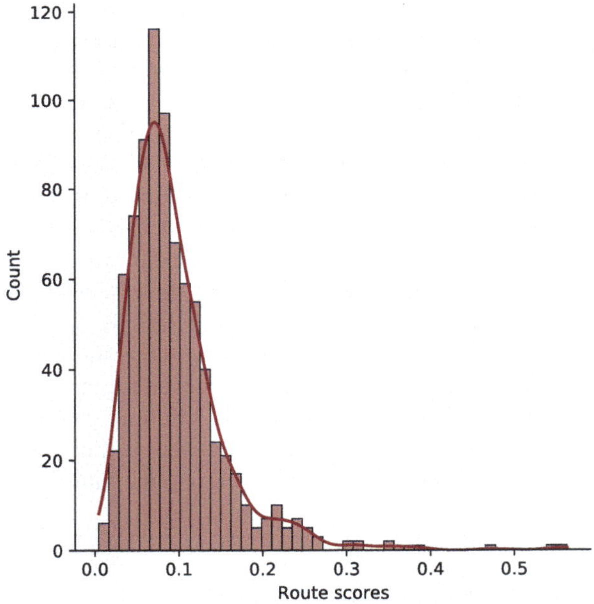

scores of predicted route sequences in the testing sample, while Table 1.1 shows score summary statistics.

It can be seen that the algorithm was capable of predicting some route sequences with high accuracy, with a score as low as 0.00447 and a maximum score of 0.563, which is still lower than a random predictor.

Table 1.1 Score summary statistics in the holdout sample with 815 routes

Statistic	Value
Mean	0.0958
Median	0.0823
Minimum	0.00447
Maximum	0.563
Std dev	0.0604

1.5 Conclusions

We have developed an approach to predict route sequences based on inverse reinforcement learning and Bayesian optimization. Unlike the approach adopted in VRPs, in which one is interested in designing optimal routes, we were interested in developing a machine learning or statistical model which may be capable of predicting routes similar to the ones generated by human drivers. We used a parameterized linear model in the cost function and tried to learn its parameters in order to minimize the loss function defined as the average prediction score.

We illustrated the proposed approach with the use of real data and could obtain predicted routes considerably better than a random predictor when compared with real routes generated by human drivers. This suggests that it may be possible to develop models which can learn local information available to drivers and which is reflected in the real routes they build. In some applications, especially in urban environments where the "last mile problem" often arises, routes planned by a machine learning model may be more feasible than routes planned by an optimization model.

We think the main limitation of our approach is that the loss function is computationally expensive and non-differentiable, so that it is hard to use more powerful approximation architectures such as artificial neural networks. We resorted to Bayesian optimization to circumvent the non-differentiability. On the other hand, it is worth noting that we used simple features in the linear model. We think better engineered features may potentially improve the prediction score. In addition, further research may allow the use of differentiable models such as neural networks.

Acknowledgments We thank Prof. Dr. Tiberius de Oliveira e Bonates, from the Department of Statistics and Applied Mathematics at the Federal University of Ceará, Fortaleza, Brazil, for reviewing the manuscript of this chapter.

References

1. P. Abbeel, A.Y. Ng, Apprenticeship learning via inverse reinforcement learning, in *Proceedings of the Twenty-First International Conference on Machine Learning* (2004), p. 1
2. D.L. Applegate, R.E. Bixby, V. Chvátal, W.J. Cook, *The Traveling Salesman Problem: A Computational Study (Princeton Series in Applied Mathematics)* (Princeton University, USA, 2007)

3. S. Arora, P. Doshi, A survey of inverse reinforcement learning: Challenges, methods and progress. Artif Intell **297**, 103500 (2021)
4. R. Bellman, *Dynamic Programming* (Princeton University, Princeton, 1959)
5. R. Bellman, Dynamic programming treatment of the travelling salesman problem. J. ACM **9**(1), 61–63 (1962)
6. S. Belogolovsky, P. Korsunsky, S. Mannor, C. Tessler, T. Zahavy, Inverse reinforcement learning in contextual MDPs. Mach. Learn. **110**(9), 1–40 (2021)
7. D. Bertsekas, D. Castanon, Rollout algorithms for stochastic scheduling problems. J. Heuristics **5**(1), 89–108 (1999)
8. D. Bertsekas, J. Tsitsiklis, C. Wu, Rollout algorithms for combinatorial optimization. J. Heuristics **3**(3), 245–262 (1997)
9. D.P. Bertsekas, *Dynamic Programming and Optimal Control*, vol. II, 3rd edn. (Athena Scientific, New York, 2011)
10. C.B. Browne, E. Powley, D. Whitehouse, S.M. Lucas, P.I. Cowling, P. Rohlfshagen, S. Tavener, D. Perez, S. Samothrakis, S. Colton, A survey of Monte Carlo tree search methods. IEEE Trans. Comput. Intell. AI Games **4**(1), 1–43 (2012)
11. T. Cazenave, F. Teytaud, Application of the nested rollout policy adaptation algorithm to the traveling salesman problem with time windows, in *International Conference on Learning and Intelligent Optimization* (Springer, Berlin, 2012), pp. 42–54
12. P.I. Frazier, Bayesian optimization, in *Recent Advances in Optimization and Modeling of Contemporary Problems* (INFORMS, New York, 2018), pp. 255–278
13. A. Gosavi, *Simulation-Based Optimization*, 2nd edn. (Wiley, New York, 2003)
14. F. Guerriero, M. Mancini, R. Musmanno, New rollout algorithms for combinatorial optimization problems. Optim. Methods Softw. **17**(4), 627–654 (2002)
15. M.. Held, R.M. Karp, A dynamic programming approach to sequencing problems. J. Soc. Ind. Appl. Math. **10**(1), 196–210 (1962)
16. D.R. Jones, M. Schonlau, W.J. Welch, Efficient global optimization of expensive black-box functions. J. Global Optim. **13**(4), 455–492 (1998)
17. A.Y. Ng, S.J. Russell, Algorithms for inverse reinforcement learning, in *Proceedings of the Seventeenth International Conference on Machine Learning*, ICML '00 (Morgan Kaufmann Publishers Inc., San Francisco, 2000), pp. 663–670
18. F. Nogueira, *Bayesian Optimization: Open Source Constrained Global Optimization Tool for Python* (2014)
19. A.R. Pitombeira-Neto, H.P. Santos, T.L. C. Silva, J.A.F. de Macedo, Trajectory modeling via random utility inverse reinforcement learning (2021). https://doi.org/10.1016/j.ins.2024.120128
20. W.B. Powell, *Approximate Dynamic Programming: Solving the Curses of Dimensionality*, 2nd edn. (Wiley, New York, 2011)
21. R.S. Sutton, A.G. Barto, *Reinforcement Learning: An Introduction*, 2nd edn. (MIT Press, New York, 2018)
22. P. Toth, D. Vigo, *The Vehicle Routing Problem* (SIAM, New York, 2002)
23. M. West, J. Harrison, *Bayesian Forecasting and Dynamic Models*, 2nd edn. (Springer, Berlin, 1997)

Chapter 2
A Comparative Evaluation of Monolithic and Microservice Architectures for Load Profiling Services in Smart Grids

Artur F. S. Veloso, José V. R. Júnior, Matheus M. do N. Costa, Ricardo A. L. Rabelo, and Placido R. Pinheiro

Abstract In recent years, the need for an efficient and scalable framework to handle the increasing demand for energy consumption data from residential consumers in the context of smart grids (SG) has been growing. In addition, this data is increasingly massive and in large volumes, making analysis slow and decision making by the power utility impractical. In addition, there is a need to transform all historical data into profiles that determine the energy consumption pattern of consumers independently, better known as load profiling (LP). Based on this, this paper proposes a comparative analysis between microservice architecture and monolithic architecture to provide LP services in SG. As a result, the microservice architecture presented significant advantages over the monolithic architecture, standing out for the modularity, independence, and scalability offered. The adoption of microservices was recommended by this work because it allows better maintainability, evolution, and flexibility in the implementation of LP services in SGs, meeting the growing demands of data and promoting greater agility in the development of these services, besides presenting a lower cost of deployment and maintenance. However, this work contributed to decision making in the scope of system architecture for the SG scenario.

Keywords Architecture · Microservices · Monolithic · Load profiling · Intelligent networks · Efficiency · Scalability

---

A. F. S. Veloso · J. V. R. Júnior · R. A. L. Rabelo
Department of Computing, Federal University of Piauí (UFPI), Teresina, PI, Brazil
e-mail: arturfdasveloso@ufpi.edu.br; valdemirreis@ufpi.edu.br; ricardoalr@ufpi.edu.br

M. M. do N. Costa
Department of Software Engineering, iCEV Faculty (iCEV), Teresina, PI, Brazil
e-mail: matheus.costa@somosicev.com

P. R. Pinheiro (✉)
Department of Computer Science, University of Fortaleza (UNIFOR), Fortaleza, CE, Brazil
e-mail: placido@unifor.br

C. A. S. Oliveira, M. P. Pardalos (eds.), *Handbook of Artificial Intelligence and Data Sciences for Routing Problems*, Springer Optimization and Its Applications 219, https://doi.org/10.1007/978-3-031-78262-6_2

2.1 Introduction

The smart grid (SG) is an advanced electrical infrastructure system that uses information and communication technologies to provide a more efficient, reliable, and sustainable management of electricity distribution. This integrated network enables two-way communication between consumers, energy suppliers, and infrastructure, allowing real-time monitoring, remote control, and optimization of energy use [10]. The SG is composed of several components, such as smart meters (SM), sensors, automation devices, data management systems, and advanced analysis algorithms [14]. This innovative approach aims to improve energy efficiency, reduce transmission and distribution losses, facilitate the integration of renewable energy sources, and provide greater interaction and engagement of consumers in managing their energy consumption.

The implementation of the SG brings numerous benefits to the electricity sector [12]. Besides promoting greater operational efficiency and cost reduction, this technology provides greater system resilience, capable of adapting to adverse conditions, such as failures and natural disasters. In addition, SG allows the integration of renewable energy sources, such as solar and wind power, in a more efficient and reliable way, contributing to the reduction of greenhouse gas emissions and environmental sustainability. Additionally, this intelligent infrastructure encourages consumer engagement, allowing better management of their energy consumption through detailed information and real-time feedback, leading to increased awareness and adoption of more sustainable consumption practices [7].

However, load profiling (LP) is born and plays a key role in SG by providing detailed information about users' energy consumption patterns. This process involves the collection, analysis, and modeling of energy consumption data at different time intervals [14]. LP helps to understand and predict energy demand, enabling more efficient planning and better management of the power grid. With LP, you can identify times of peak demand, detect anomalies or deviations in consumption, and optimize the sizing and allocation of energy resources. In addition, LP enables the implementation of demand control strategies, such as demand response and differentiated tariffs, which encourage consumption moderation and the adoption of more sustainable practices. In summary, the LP plays a crucial role in the SG, providing valuable information for operational efficiency, demand forecasting, and promoting smarter and more conscious energy consumption [16].

However, the effective application of LP services in SGs faces significant challenges, especially regarding scalability and software architecture [6]. One of the main problems is related to the massive collection of energy consumption data, since SGs deal with a large number of connected devices, generating a huge amount of information. Handling this amount of data efficiently requires scalable software architectures capable of processing, storing, and analyzing the load data in real time [3]. In addition, interoperability and system integration also become challenges, as common standards for communication and data sharing between

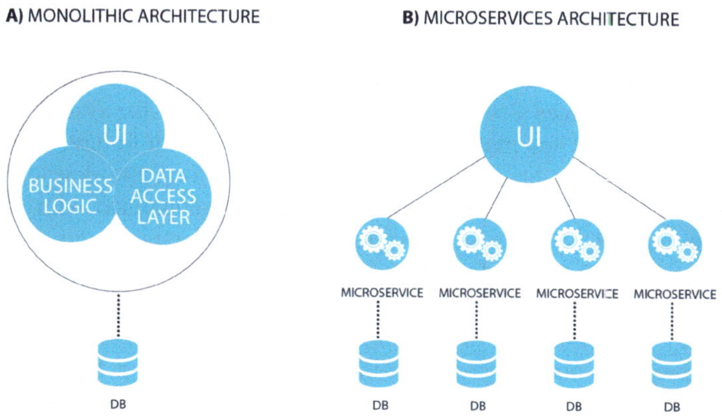

Fig. 2.1 Monolithic versus microservice architecture

different entities in the SG need to be established. The software architecture needs to be designed in a flexible and modular way, allowing the smooth integration of different components and systems. Ensuring data security and privacy is also a key concern, as sensitive information about energy consumption is collected and processed. Therefore, addressing these software architecture and scalability challenges is essential for LP services to be applied effectively in SGs, enabling more efficient and sustainable SG management [1].

Based on this, it can be noted that to have a scalable and robust infrastructure capable of meeting the execution needs of this range of data and services, a new problem is highlighted, the use of monolithic software architecture, which in turn has been common in the development of software systems, especially in enterprise and commercial applications with low execution power and scalability [22]. A monolithic architecture is characterized by being a complete, integrated system, where all functionality is developed in a single package, as shown in Fig. 2.1. While the (a) monolithic architecture may be easy to develop and deploy initially, it can lead to scalability and maintenance problems in the long term [2]. One of the main problems with monolithic architecture is its lack of modularity. This means that as the system grows and evolves, changes to one feature can impact other parts of the system. In addition, scalability can also be a challenge in monolithic architectures, as increasing workloads can lead to performance issues.

To overcome these challenges, many works are adopting the microservice architecture [9, 13]. (b) Microservice architecture is an approach to software development that divides the system into small independent services, each responsible for a specific functionality, as shown in Fig. 2.1. These services are deployed and run separately, which allows them to be scaled and updated independently of each other. This architecture approach has been widely adopted by works seeking flexibility and scalability in their systems [9]. By adopting the microservice architecture, it is possible to create a highly modular and scalable system where each service can be developed, tested, and deployed independently. This allows services to act in different parts of the system at the same time, without interfering with other services. In addition, microservices also help increase the resiliency of the system, since each service can be isolated and managed independently in case of failure.

Therefore, this work has as main objectives to perform the development and comparative analysis of microservice architecture in comparison with monolithic architecture for LP services in SGs. It seeks to understand and evaluate the advantages and challenges of each approach, taking into account criteria such as modularity, independence, scalability, and maintainability. This work aims to contribute to decision making in choosing the most appropriate architecture to deal with the growing demand for data in this context, promoting better operational efficiency, flexibility, and agility in the development and maintenance of LP services in SGs. Furthermore, it is intended to provide valuable insights for architects and developers involved in the design and implementation of SG systems, driving the technological advancement and improvement of the energy infrastructure.

In addition, presents where the application will act and how will be the process of collecting data from the communication infrastructure a hybrid communication infrastructure with LoRaWAN and LoraMesh for the demand side management as a service (HyDSMaaS) from [23]. This work will be implemented using the Amazon cloud (AWS) and with a real data package with energy consumption data collection every 30 minutes with over 5500 consumers in a city. In the end, the results will be presented based on a real experiment, to test definitively.

Additionally, this paper aims to answer the following scientific questions:

1. Which software architecture is best suited to enable the scalable and robust deployment of SG services for the SG scenario?
2. Which architecture is best suited for the proposed SG scenario, monolithic or microservice?

Additionally, this paper presents the following contributions:

- Architecture comparison: The work provides a detailed comparative analysis between microservice architecture and monolithic architecture in the context of LP services in SGs. This comparison provides a deeper understanding of the advantages and challenges of each approach, providing a solid foundation for future research and projects in the area.
- Evaluation of relevant criteria: The study considers important criteria, such as modularity, independence, scalability, and maintainability, in the analysis of the architectures. This evaluation contributes to a better understanding of the key factors to be considered when selecting the appropriate architecture for LP services. It also helps identify gaps and areas for improvement in existing approaches, providing valuable guidelines for future system development in this area.
- Practical insights for practitioners: The work aims to provide practical insights for architects and developers involved in the design and implementation of SG

systems. The analyses and results obtained provide valuable information for making informed decisions in developing efficient architectures and solutions. This contributes to the effective application of LP services, promoting a smarter, more sustainable, and efficient energy infrastructure.

Finally, this paper is organized as follows: in Sect. 2.2 is a study of the current state of the art with respect to monolithic versus microservice architecture. The practical methodology of this work is detailed in Sect. 2.3. The results are presented in Sect. 2.4. And finally, Sect. 2.5 presents the conclusions and future works.

2.2 Related Works

The context of microservices and monolithic architectures for SG has been focused on exploring the advantages and challenges of each approach [11]. Several studies have focused on the development of microservice architectures for SG, aiming to improve the modularity, scalability, and flexibility of systems [11]. These architectures allow the implementation of independent services, making it easy to add, remove, and update components without affecting the entire system. Furthermore, orchestration and microservice management techniques are explored to ensure efficient coordination of the different parts of the system.

Regarding the monolithic architecture, studies have investigated ways to optimize its performance and simplify its maintenance. Approaches such as internal modularization and the use of design patterns have been explored to improve the logical division and separation of responsibilities within the monolithic architecture. In addition, virtualization and containerization techniques are applied to facilitate the deployment and scalability of these systems [18].

Overall, research is focused on finding solutions that meet specific SG requirements such as high availability, security, and energy efficiency [4]. In addition, there is growing interest in issues related to the integration of heterogeneous systems, large-scale data management, and cybersecurity. These studies aim to provide guidelines and practical solutions for developing efficient and reliable architectures for implementing SG services, driving technological advancement in this critical area.

To this end, the monolithic architecture is analyzed in [22] where the authors raise that its main advantages are easy implementation, centralized design, and fast to implement; however, the monolithic architecture also has significant problems. The main disadvantage is that adding new features or functionality can be difficult and time-consuming as it requires modifications to the entire code base. In addition, scalability can be limited as all instances of the application must be deployed together as raised in the work [2].

On the other hand, the microservice architecture allows the different components of the application to be developed and deployed as independent services, making the

Table 2.1 Work that uses monolithic and microservice architecture for SG

Work	Architecture	Scalability	Cost
[2, 8, 17, 22]	Monolithic	Low	High
[5, 9, 13, 21]	Microservice	High	Middle

Table 2.2 Comparison between related works and the proposal of this work

Work	Architecture	Scalability and cost
This work	Microservice	Generating and clustering LP
[19]	Monolithic	Smart plug
[22]	Microservice	Large codebases
[8]	Monolithic	Green energy management

addition of new features and functionality easier and faster. In addition, scalability is easier to manage because each service can be deployed and scaled independently, as the authors talk about in [9]. However, microservice architecture also presents significant challenges, as presented in the paper [13]. The complexity of managing many independent services can be high, and communication between services can be a source of problems. In addition, implementing a microservice architecture can be more time-consuming and expensive than a monolithic architecture. Additionally, this paper has listed some of the major works that have used the architectures as a fundamental basis in their development; these works are listed in Table 2.1.

As presented in Table 2.1, the monolithic architecture mostly presents low scalability and high deployment cost. While in the microservice architecture, the scalability is high and the cost is lower. For this reason, the architecture used in this work will be the microservice architecture, having as its main objective the scalability and low cost of deployment. To conclude, this work will evaluate the cost of deployment in both architectures in the proposed scenario, to highlight the value of the cost of deployment and maintenance, further validating the proposal of this work.

Finally, in Table 2.2 are all the main related works and the proposal of each of them, so that it can be possible where this work intends to act and contribute.

Therefore, this paper proposes the development and comparative analysis of microservice architecture versus monolithic architecture for LP services in SGs. The difference from the state of the art lies in the specific approach to the analysis and comparison of these architectures, with a focus on the context of LP services. While many studies address the architectures more generally, this work focuses on understanding the advantages and challenges of these specific approaches to address the increasing data demand and operational efficiency requirements in this specific context. Furthermore, the work seeks to provide practical insights for architects and developers in choosing the most appropriate architecture, considering relevant criteria such as modularity, independence, scalability, and maintainability. With this approach, we hope to contribute to the advancement of knowledge and provide more targeted guidance for the efficient implementation of LP services in SGs.

2.3 Microservice Versus Monolithic Architecture for Load Profile Services for Smart Grid

This work has as its main objective the implementation, evaluation, and proposal of an architecture that is scalable and robust and that can meet all the demand both of processing and storage, as well as the demand that the SG scenario requires. To this end, this work desolved a microservice architecture that is divided and described in five steps, where the first is the generation of the LP through a pre-processing based on time series, then the microservice architecture developed is presented, and the cost value of the microservice architecture versus monolithic and analysis of clustering algorithms both classical and deep clustering for the proposed scenario are presented, and then the clustering of the LPs in groups of similar or close profiles with the algorithm that obtains the best result is performed, and finally new services based on real data from the city of London with more than 5500 consumers are adopted https://data.london.gov.uk/dataset.

2.3.1 Load Profile Services

The LP of electricity is a graphical representation of the amount of energy consumed in a given period of time by a final consumer, be it a residence, a company, or a building. The LP can be used to understand and forecast electricity consumption in a specific geographical or time scenario, in order to analyze usage patterns and peak times and even suggest new ways to use energy consumption or predict consumption in the coming days, months, or years. However, to identify an LP based on the energy consumption data of a consumer, one must first have a database of at least 1 month, with readings at least every 30 minutes. With the database in hand, the following steps should be followed:

1. Collect energy consumption data by the hour or every 30 minutes: This can be done by measuring energy consumption at a particular location, using an SM, or by collecting data from an external source, such as an energy service provider.
2. Plot the raw data collected every hour or 30 minutes for at least a month: Creating a scatter plot representing all consumption for a given period, by the hour, can be plotted every hour or every half hour.
3. Identify patterns using the time series model: A time series model is a statistical approach that allows to analyze and predict the behavior of data over time. To identify energy consumption patterns using this model, it is necessary to have a historical series of data representing energy consumption at regular intervals of 1 hour or every 30 minutes. Based on this data, has been using time series analysis techniques, such as seasonal decomposition, to identify trends and seasonal patterns in energy consumption. From this, future forecasts of energy consumption can be made based on the patterns identified and other

relevant factors such as climate change or holidays. These forecasts can help improve energy efficiency and reduce costs for energy consumers and suppliers. In addition, it is also possible to generate a pattern per hour, which makes it possible to create an LP pattern from the historical data collected by the Meter Data Management System (MDMS).

4. Plot the LP data from after performing the time series model; it will be possible to generate the consumption per hour, where 48 energy consumption records will be plotted for each 30-minute hour, from 00:00 to 23:30, as this paper intends to work.

2.3.2 Microservice Versus Monolithic Architecture

Software architectures can be categorized in many ways, but two of the most common are microservice and monolithic architectures. Monolithic architecture is a traditional approach in which all software is built as a single block of code with all functions integrated. The microservice architecture, on the other hand, consists of dividing the software into smaller, independent parts, called services, each with its own business logic and well-defined responsibilities. Both architectures have their advantages and disadvantages and can be applied in different scenarios, depending on the project needs. The monolithic architecture is best suited for smaller projects, where simplicity is more important than scalability. It is also easier to implement and maintain, since all the code is in one place. On the other hand, the microservice architecture is best suited to larger, complex projects where scalability and flexibility are essential. It allows different services to be developed and updated independently without affecting the rest of the system.

In short, the choice between a monolithic architecture and a microservice architecture will depend on the specific requirements of the project and the balance between simplicity and scalability. Both have their advantages and disadvantages and can be successfully applied in real-world scenarios, each with its own way of contributing. With this in mind, this work decided to apply both architectures in the proposed scenario, which is the SG scenario, where applications must have high availability, scalability, processing power, and robustness and be able to meet a big data latency demand, which has a large movement every half hour.

To do this, some of the main parameters were defined to define which architecture to use:

- Scalability: The microservice architecture is designed to scale horizontally, allowing different parts of the system to scale independently. Monolithic architecture, on the other hand, tends to be more difficult to scale.
- Flexibility: Microservice architecture allows parts of the system to be replaced or upgraded independently, while monolithic architecture is generally less flexible.

- Response time: Microservices can respond more quickly to requests because the parts of the system are smaller and more specialized. The monolithic architecture may have slower response times due to the size and complexity of the system.
- Maintainability: Microservice architecture is generally easier to maintain because the parts of the system are smaller and more specialized. Monolithic architecture can be more difficult to maintain due to the size and complexity of the system.
- Complexity: The microservice architecture tends to be less complex than the monolithic architecture because the parts of the system are smaller and more specialized.
- Performance: The microservice architecture, on the other hand, is better at handling high-intensity workloads, while the monolithic architecture may be more efficient for less intense workloads.
- Monitoring and tracking: The microservice architecture has more capabilities to track and monitor the performance of services, while the monolithic architecture can be more difficult to monitor and track.
- Resource optimization: The microservice architecture is better at resource optimization because services are independent and can scale independently, while the monolithic architecture does not have the same flexibility.

Therefore, both architectures were modeled so that this analysis can be done with more precision and is rich in information. With this, the monolithic architecture was modeled as presented in Fig. 2.2, which shows the cloud infrastructure so that the services can act. Initially, The Things Networking (TTN) will be responsible for grouping the data and sending it to the MDMS service (EC2 C5G.8XLARGE),

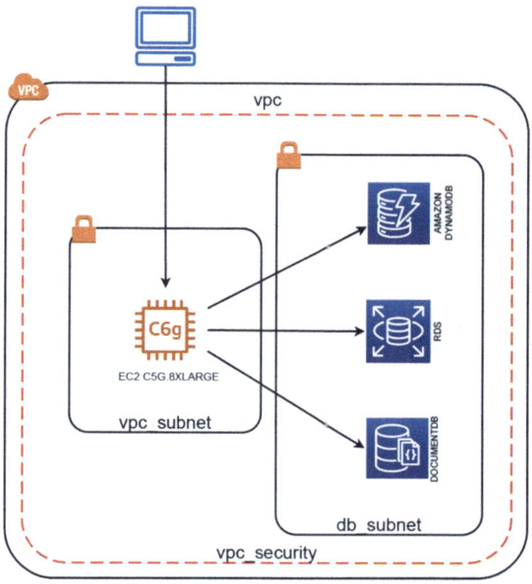

Fig. 2.2 Monolithic architecture for the proposed scenario

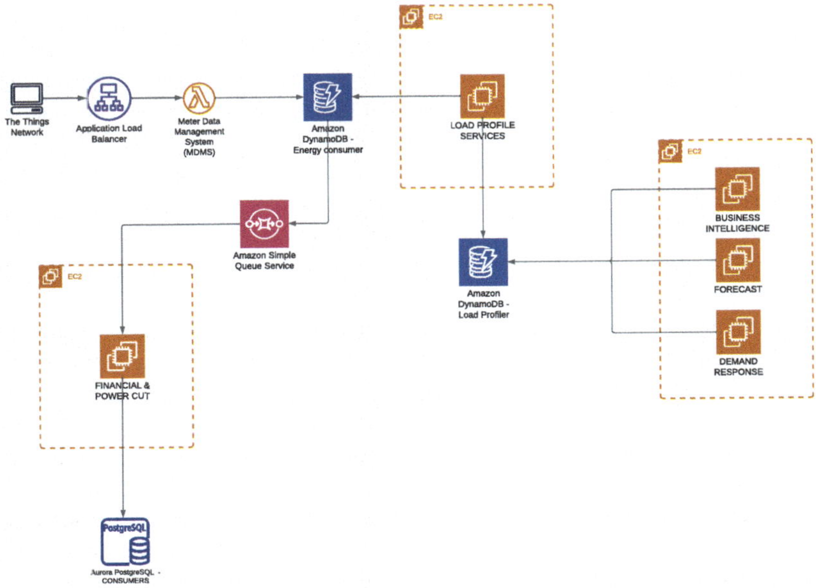

Fig. 2.3 Microservice architecture for the proposed scenario

which will be responsible for receiving the raw data every 30 minutes from all consumers and storing it in a general database. Besides this service, all the services that will be implemented in this architecture will run in the same place, which is an Amazon Web Service (AWS) EC2 instance [24] that is prepared to process all these services, i.e., it is a robust machine with configurations that allows meeting the great demand that this architecture requires. However, this architecture can overload, if some or all of the services need a higher demand than this infrastructure allows. Another negative point of this architecture is that if this instance becomes slow because of some of the services, all the others will be affected, causing the unavailability of some of the systems.

On the other hand, the microservice architecture was modeled in Fig. 2.3, where the TTN also has the important role of grouping the data and sending it to the MDMS that does the same role of storing all the data in the general database. However, in this architecture, before the MDMS service, there is a load balance application that plays the role of managing the processing supply based on the demand that is generated by the TTN, that is, it is responsible for distributing the data to each function that is created by the AWS serverless lambda function, which self-adapts to the demand generated, and is where the MDMS is being executed. The general database is in AWS, but in a serverless DynamoDB, a self-scalable NoSQL (non-relational) database [15] service is also from AWS. From there, the other services are installed on separate EC2-configured instances of AWS, where as new services are implemented, new instances are created to serve each application independently and separately. This way, if one instance suffers some

Table 2.3 Microservice architecture processing data

Bytes	Time (ms)	Packet count	Total time (ms)
63	75.18	1	75.18
63	75.18	2,60,000	up to 300

Table 2.4 Monolithic architecture processing data

Bytes	Time (ms)	Packet count	Total time (ms)
63	75.18	1	75.18
63	75.18	2,60,000	up to 19546800

kind of problem or slowdown, it does not affect the instance of another service, thus leaving the architecture scalable, resilient, and with high availability.

Additionally, a queue service was implemented that will be responsible for managing the queue of the financial and cut control service, that is, as this service is executed, it consumes a queue, thus generating a way to execute the service without affecting the database instance, leaving it scalable and resilient.

With the architecture, in Table 2.3, the configurations of the scenario of this work will be defined. Initially, a scenario of 2,60,000 consumers will be considered, where each consumer will send a data packet with size of 63 bytes, as defined and detailed in the work [23], with all the necessary data for the LP application. Each consumer will send a packet with its consumption data every 30 minutes, with an average time of 75.18 milliseconds (ms) each. As there are 2,60,000 consumers in this example, because we are using the lambda function [20], it will try to process all 2,60,000 packets at the same time, and it may happen that one or the other will have some kind of delay; this way, AWS guarantees a processing of up to 300 ms for the 2,60,000 data packets in this scenario.

In Table 2.4 are the processing data of the monolithic architecture, where the processing is done in a non-scalable way, that is, one request at a time. Since 2,60,000 packets are sent every 30 minutes, the total execution time to process them is 19546800 ms, or more than 5 hours. This makes its adoption in the real 350 scenarios impossible; with the processing time is longer than this, making its adoption impossible.

When it comes to the financial part, this work also budgeted the monthly cost value of both architectures; in Table 2.5 is the cost value of the monolithic architecture, making a total of $1720.72 per month. It is worth noting that the infrastructure of AWS is smaller than in the microservice, but the configuration is superior, because it needs to meet the same demand. The server is an EC2 model c6g.8xlarge 32 vCpu 64 gb, while the database server is AWS RDS (relational database service), where a NoSQL database in DynamoDB and a mongo Document DB server, also from AWS, were used.

On the other hand, in Table 2.6, the cost value per month of the microservice architecture is U$1394.96, being a larger configuration, but scalable, robust, and resilient, which makes its configuration inferior and the cost considerably lower, thus making it financially more viable.

Table 2.5 Monolithic architecture costs per month

Service	Cost (US$)	Description
EC2	794.24	c6g.8xlarge 32 vCpu 64 gb
RDS	259.88	
DynamoDB	474.50	
Document DB	192.10	

Table 2.6 Cost of microservice architecture per month

Service	Cost (US$)	Description	Details
Load Balancer	0.48	0.03276 GB per hour	
Lambda	187.68	520000 Requests per hour	Processing 7117500 GBs
EC2	280.32	m6i.2xlarge 8 vCpu 32 gb	3.5 GHZ
SQS	0.00	1.000.000,00 total of standard queue requests	
RDS	259.88		
DynamoDB	474.50		
Document DB	192.10		

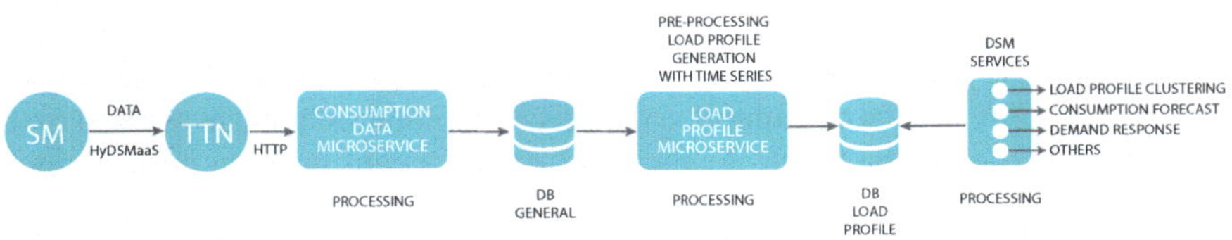

Fig. 2.4 Architecture of the microservice steps for the proposed scenario

To conclude, this work will use the microservice architecture because it presents a better and more prepared infrastructure for the proposed scenario, as well as its cost is more viable than the monolithic architecture.

With this in mind, the microservice architecture proposed by this work was implemented as shown in Fig. 2.4. The SMs of all consumers send the data packets to the TTN, which will play the role of managing the reception of the packets and forwarding them to the MDMS microservice, which is responsible for receiving the consumption data and storing it in the general database. From then on the other microservices will act on top of this general database; however, a pre-processing microservice was implemented, which is responsible for processing the raw data, generating the LP of all consumers, and storing in a second database only with LP data of each consumer. In this case, the size of the database to be used by the new microservices that will be implemented in the future is significantly reduced, such as LP clustering services, energy consumption and demand forecasting, demand response, and other services.

For implementation and validation of the proposal of this work, a database was used, considered big data, which represents the consumption data of 5562 consumers in London. The readings were collected between November 2011 and February 2014, and the original data file has about 11 GB subdivided into 167

CSV files with about 167 million records. Among the columns are: Consumption (kilowatt-hour—KWh), DateTime (every half hour), LCDID (consumer code), StdorToU (if the consumer is subject to the dynamic time-of-use tariff—dToU), and dToU (tariff). This database is publicly available at https://data.london.gov.uk/dataset.

To start the implementations, 5562 consumers were registered in the application, and 167 million records were registered by the MDMS service in the global database, thus storing the raw data. Once this was done, the LP generation microservice was executed, resulting in a generation of 5562 LPs, stored in the LP database. With this, we have the first result, which is the size of the data storage, reducing from 167 million records to 5562 records. The other results are described in the next section.

2.4 Results and Discussion

During the comparison between the two architectures, to configure in both environments an architecture that meets the needs of the proposal of this work, the monolithic architecture demanded a value of U$1720.72 per month, while the microservice architecture demanded a value of U$1394.96 per month, obtaining a lower value. It is worth noting that the microservice architecture configuration is more robust and has more components; however, the configurations are simpler, because in this architecture the services are distributed, while in the monolithic architecture, they are all executed in a single instance.

Finally, the decisive fact in choosing the microservice architecture was the execution time. While it can execute 2,60,000 records in up to 300 ms, the monolithic architecture takes about 5 hours, and this time makes its use unfeasible because the execution time cannot exceed 30 minutes, which is the interval time between the 2,60,000 dispatches every 30 minutes during the day, totaling 48 dispatches.

Figure 2.5 shows the result where the stress tool automatically balanced the bots that would send the payloads (limited to 100 bots). This way the tool identified that the monolithic would be better to do the firing in a more gradual way, while in the microserver it was possible to have a greater "explosion" in the firing, that is, the microserver in this case was more resilient to receiving payloads from the bots.

Meanwhile, in Fig. 2.6 the response time in the monolithic is limited to the demand and the power of the machine, and the microswitches are running on a serverless system that for this reason manages to keep the response time low, even though it has more oscillation than the monolithic. In other words, the monolithic system has a problem, because there comes a time when the demand is higher than it can support and this is when the response time can take longer or even crash, while in the microserver, because it is being executed in a serverless system, it has the resilience to adapt to demand, that is, when it has little demand it will execute little processing, and when it has a lot of demand, it will have a lot of processing

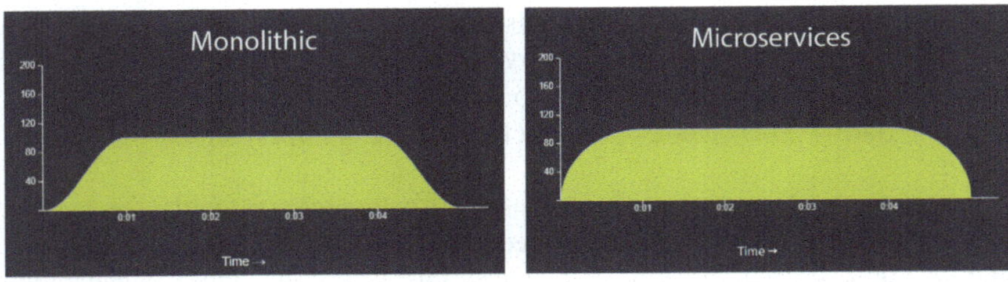

Fig. 2.5 Planned bot count

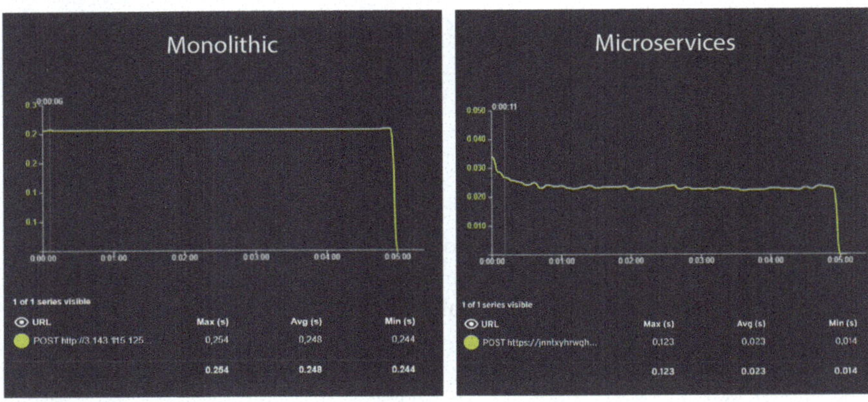

Fig. 2.6 Average load time by URL

with minimal risk of crashing or slowdown. However, the time of the microservice excels compared to the monolithic one.

It is worth noting that the average time in the monolithic was 0.248 seconds, while in the microservice it was 0.023 seconds, a significant time better than the monolithic, especially thinking in the long term and on a larger scale.

The percentage of response time was used to divide the processing and understand at what point the processing did or did not require a longer or shorter processing time. For this, Fig. 2.7 presents the execution of the two architectures in percentile, showing that at all times the monolithic is in maximum processing, while in the microservice, in some smaller response times the execution is low, and in some moments where the demand increases, the maximum execution is requested, hence the importance of this architecture; the flexibility helps this to happen.

The network throughput was analyzed in both architectures, and the results are presented in Fig. 2.8. In it, the upload of data that was made by the bots, there are differences because the amount of requests was different in the monolithic and microservices; in the download the difference was high because of the payload that the API sent. In the monolithic the response payload had less data, because of how the platform was running and how the script was adapted, while in the microserver the response payload contained more data because the serverless system added more

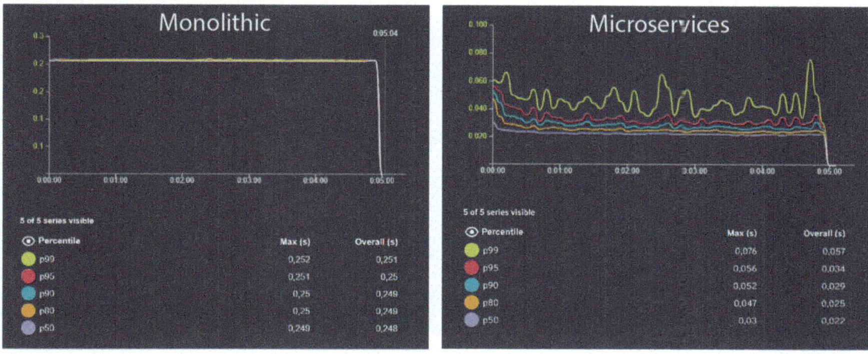

Fig. 2.7 Response time percentiles

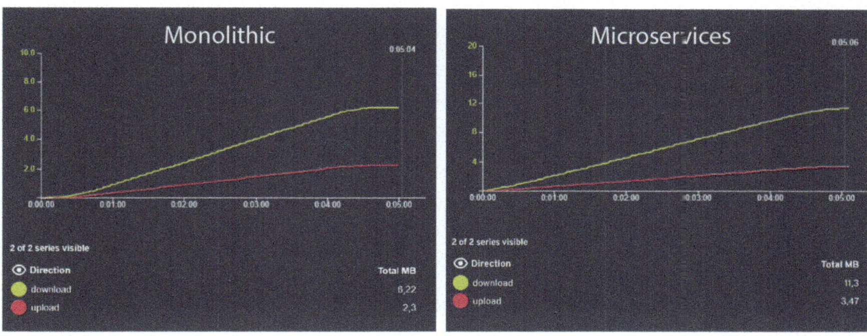

Fig. 2.8 Cumulative network throughput

information in the response header. Finally, it can be concluded that the microservice network performance was higher than the monolithic one, practically doubling.

Additionally, the relations between requests per time were analyzed; for this, in Fig. 2.9 the comparison of hits on the endpoint is presented, as well as the quantities of requests per second. In this context, pages and hits have the same value because they were tested only with a single endpoint; in the microserver there were errors that were handled by the serverless system itself. Overall the monolithic had an average of 34.9 requests per second and maximum of 45.3 requests per second, while in the microservices the results were better, with an average of 43.7 and maximum of 49.8 requests per second, a significant and expressive result.

Finally, it was analyzing the percentage of CPU execution of the servers to receive the packets sent by the SMs; with this, in Fig. 2.10 presents the results in the monolithic architecture, which in turn obtained the average usage of 7.4%, while to make the sending for microservice architecture the consumption was 8.7%. However, less CPU usage is required in the monolithic architecture than in the microservice, but the difference of 1.7% is not considerable, considering the other results seen previously.

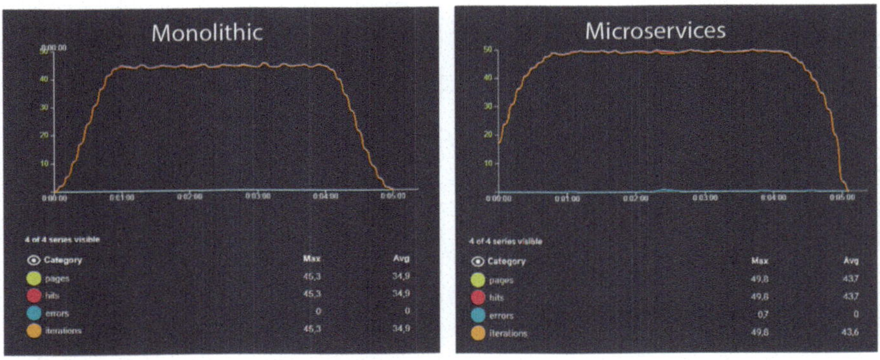

Fig. 2.9 Transaction throughput

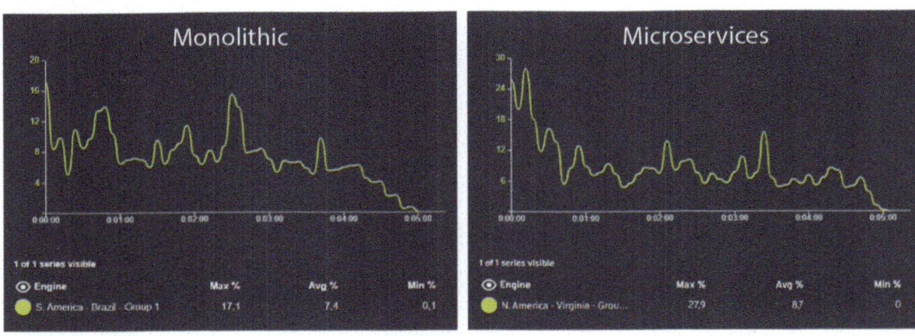

Fig. 2.10 Engine CPU utilization

Therefore, the use of microservice architecture is considered to be the most appropriate to enable the scalable and robust implementation of SG services for the SG scenario. This is because the microservice architecture allows the separation of services into independent units, which can be individually scaled and updated without affecting the operation of other services. In addition, the microservice architecture allows the use of different technologies and tools for each service, according to their specific needs. In this way, it is possible to create highly specialized and personalized services, capable of meeting the demands of the customers and the electrical system as a whole more efficiently. The microservice architecture also facilitates integration with other systems and platforms, allowing the creation of more open and flexible solutions, which is fundamental for the SG scenario, which is characterized by a wide variety of power generation sources and distributed consumption devices.

Additionally, the microservice architecture is the most appropriate for the SG scenario because it allows the creation of specialized and customized services, capable of meeting the specific demands of customers and the electric system as a whole. This is because the microservice architecture allows the separation of services into independent units, which can be scaled and updated individually, without affecting the operation of other services. In this way, it is possible to create services that more efficiently meet customer needs, such as the creation of incentive programs to reduce electricity consumption during peak hours. In addition, the microservice architecture facilitates integration with other systems and platforms, allowing the creation of more open and flexible solutions, which is fundamental for the SG scenario, which is characterized by a wide variety of power generation sources and distributed consumption devices. The monolithic architecture, on the other hand, is not as well suited for the SG scenario, as it makes it difficult to scale and update services individually, and limits the flexibility and customization of services.

During the execution of the tests, the Loadster tool was used to monitor the services; the execution time to obtain the data from the 5500 consumers was 5 minutes, yesterday in the monolithic it had 10683 iterations, and in the microservice there were 13576 iterations, that is, in the monolithic, some packets were lost. The average response time in the monolithic was 0.248 seconds, while in the microservice it was 0.023 seconds, achieving a significantly better result.

Finally, the proposal of this work was implemented and tested using a real database from London, with more than 5500 final consumers. The database contains more than 167 million records and is considered big data, since its expansion is performed every 30 minutes during the day. As an overall result, the work presented a microservice architecture that provided a new way to collect consumption data in a scalable, robust, and resilient way, so that the architecture would not overload. In addition, microservices were presented to assist SG services, such as LP generation from time series, and data clustering using classical and deep-clustering algorithms.

As main contributions, this work proposes a new deep-clustering-based microservice model for demand-side energy management as a service, furthermore develops a deep-clustering algorithm that uses unsupervised learning to identify consumers' electricity consumption patterns, proposes the use of the microservice as a service for consumers allowing them to monitor their own electricity consumption and make informed decisions about their energy use, and proposes the use of deep learning techniques for clustering electricity consumption data and the use of microservices for demand-side energy management, which enables a more agile and scalable approach to electricity management, with low implementation and maintenance costs. Furthermore, it is possible to highlight that this work is relevant to become a basis for further research that intends to study and implement new SG services, to improve, act, and assist energy management, either on the demand side or on the EPC side.

2.5 Conclusions and Future Works

The present work addressed the problem of developing and analyzing microservice architecture versus monolithic architecture for LP services in SGs. The proposal was to perform a detailed comparison between these approaches, considering relevant criteria such as modularity, independence, scalability, and maintainability. Based on the results obtained, it can be concluded that both architectures have distinct advantages and challenges. The microservice architecture proved to be more suitable for environments that require flexibility, scalability, and a greater ability to adapt to changes, because it was able to perform the SM data collection in ten times less time than the monolithic one. On the other hand, the monolithic architecture proved to be more suitable for simpler and more stable systems, where maintenance and component integration are less complex.

As presented in the tests with a limit of 100 bots, the response time in the monolithic architecture was limited, and its execution remained constant at the maximum, thus, resulting in higher operational costs and an average execution time of 0.248 seconds. In contrast, in the microservice architecture, the processing peaked during high demand and decreased during low demand, achieving an average processing time of 0.023 seconds, significantly lower than in the monolithic architecture. When it comes to network performance, the microservice practically doubled its performance, becoming more efficient and with better performance than the monolithic one. Meanwhile, in the analysis of requests per time, the monolithic had an average of 34.9 requests per second and maximum of 45.3 requests per second, while in the microservice the results were better, with an average of 43.7 and maximum of 49.8 requests per second, thus making the microservice faster and more efficient in processing times per time than the monolithic. Finally, an analysis of receiving packets from the SMs was performed; with this, the CPU consumption of the server was analyzed, where the monolithic was 7.4%, while to make the sending to the microservice was 8.7%; in this sense for the client side the monolithic requires less CPU usage to receive the data from the client. However, the execution time and the cost of this are greater, while the difference is only 1.3% for an architecture that has a lower cost and shorter execution time, not invalidating the conclusion that the microservice is the ideal architecture for the scenario proposed by this work.

The results also indicated that the choice of the appropriate architecture depends on the specific characteristics of the SG context and the requirements of the LP service. It is crucial to consider factors such as system complexity, data demand, security requirements, and maintainability. Furthermore, it is important to note that the comparative analysis performed in this paper provides a solid basis to guide decision making, but each implementation must be evaluated individually, taking into account the specific peculiarities and needs of each project.

Overall, this work contributes to the advancement of knowledge in the area of architectures for LP services in SGs. The results obtained provide a clear view of the advantages and challenges of each approach, assisting architects and developers in selecting the most suitable architecture for their specific needs. This deeper understanding is essential to promote an efficient implementation of LP services, contributing to the improvement of operational efficiency, flexibility, and agility in SG management.

Science questions:

1. What is the most suitable software architecture to enable the scalable and robust implementation of SG services for the SG scenario? The architecture developed contemplates the use of microservices with high availability, using load balance and lambda function to improve application resiliency.
2. Which architecture is best suited for the proposed SG scenario, monolithic or microservice? As an answer, the microservice architecture was the one that obtained the best result and became the most suitable for this scenario.

There are several possibilities for future work related to the topic addressed in this paper. One is the investigation of advanced techniques for orchestration and management of microservices in the microservice architecture for LP services in SGs. This may involve the development of intelligent algorithms to optimize the coordination between different services and improve the efficiency and scalability of the system. In addition, the use of emerging technologies, such as cloud computing and the Internet of Things (IoT), can be explored to enhance the collection, processing, and analysis of real-time load data. Another possibility is to conduct further studies on the security and privacy of data in smart grid load profiling services. This includes examining encryption, access control, and anonymization techniques for energy consumption data to ensure the confidentiality and integrity of the information. In addition, one can explore approaches for secure data sharing among the different actors involved in the SG, such as energy utilities, service providers, and consumers. Finally, it is worth considering conducting studies that explore the application of machine learning and artificial intelligence techniques for the analysis of load data in SG. This may involve the development of more advanced load forecasting algorithms, the detection of anomalies in energy consumption, and the identification of consumption patterns for the optimization of grid operation. These approaches can contribute to more efficient and sustainable energy management, enabling better resource allocation and reduced environmental impacts.

References

1. K.A. Abdulsalam, J. Adebisi, M. Emezirinwune, O. Babatunde, An overview and multicriteria analysis of communication technologies for smart grid applications. e-Prime Adv. Electr. Eng. Electron. Energy **3**, 100121 (2023)
2. B. Andrade, S. Santos, A.R. Silva, From monolith to microservices: Static and dynamic analysis comparison. arXiv preprint arXiv:2204.11844 (2022)
3. M.S. Bakare, A. Abdulkarim, M. Zeeshan, A.N. Shuaibu, A comprehensive overview on demand side energy management towards smart grids: challenges, solutions, and future direction. Energy Inform. **6**(1), 1–59 (2023)
4. G. Blinowski, A. Ojdowska, A. Przybyłek, Monolithic vs. microservice architecture: a performance and scalability evaluation. IEEE Access **10**, 20357–20374 (2022)

5. B. Cen, C. Hu, Z. Cai, Z. Wu, Y. Zhang, J. Liu, Z. Su, A configuration method of computing resources for microservice-based edge computing apparatus in smart distribution transformer area. Int. J. Electr. Power Energy Syst. **138**, 107935 (2022)
6. Z. Chen, A.M. Amani, X. Yu, M. Jalili, Control and optimisation of power grids using smart meter data: a review. Sensors **23**(4), 2118 (2023)
7. M. Cho, H. Aki, D. Kodaira, Design and implementation of power management system integrated with cooperative systems for an apartment building. J. Build. Eng. **73**, 106809 (2023)
8. A. Fakhar, A.M.A. Haidar, M.O. Abdullah, N. Das, Smart grid mechanism for green energy management: a comprehensive review. Int. J. Green Energy **20**(3), 284–308 (2023)
9. M. Hamza, Transforming monolithic systems to a microservices architecture. ACM SIGSOFT Softw. Eng. Notes **48**(1), 67–69 (2023)
10. C. Ibrahim, I. Mougharbel, H.Y. Kanaan, N. Abou Daher, S. Georges, M. Saad, A review on the deployment of demand response programs with multiple aspects coexistence over smart grid platform. Renew. Sust. Energ. Rev. **162**, 112446 (2022)
11. P. Jatkiewicz, S. Okrój, Differences in performance, scalability, and cost of using microservice and monolithic architecture, in *Proceedings of the 38th ACM/SIGAPP Symposium on Applied Computing* (2023), pp. 1038–1041
12. G. Kumar, L. Kumar, S. Kumar, Multi-objective control-based home energy management system with smart energy meter. Electr. Eng. **105**(4), 2095–2105 (2023)
13. P. Mangwani, N. Mangwani, S. Motwani, Evaluation of a multitenant saas using monolithic and microservice architectures. SN Comput. Sci. **4**(2), 185 (2023)
14. T. Mazhar, R.N. Asif, M.A. Malik, M.A. Nadeem, I. Haq, M. Iqbal, M. Kamran, S. Ashraf, Electric vehicle charging system in the smart grid using different machine learning methods. Sustainability **15**(3), 2603 (2023)
15. R. Nascimento, D. Brandao, A new real-time shm system embedded on raspberry pi, in *European Workshop on Structural Health Monitoring: EWSHM 2022-Volume 1*, vol. 253 (Springer Nature, Berlin, 2022), p. 386
16. G.E. Okereke, M.C. Bali, C.N. Okwueze, E.C. Ukekwe, S.C. Echezona, C.I. Ugwu, K-means clustering of electricity consumers using time-domain features from smart meter data. J. Electr. Syst. Inf. Technol. **10**(1), 1–18 (2023)
17. M. Pau, M. Mirz, J. Dinkelbach, P. Mckeever, F. Ponci, A. Monti, A service oriented architecture for the digitalization and automation of distribution grids. IEEE Access **10**, 37050–37063 (2022)
18. P. Plecinski, N. Bokla, T. Klymkovych, M. Melnyk, W. Zabierowski, Comparison of representative microservices technologies in terms of performance for use for projects based on sensor networks. Sensors **22**(20), 7759 (2022)
19. F. Sanchez-Sutil, A. Cano-Ortega, Smart plug for monitoring and controlling electrical devices with a wireless communication system integrated in a lorawan. Expert Syst. Appl. **213**, 118976 (2023)
20. P. Sbarski, S. Kroonenburg, *Serverless Architectures on AWS: With Examples using AWS Lambda* (2017)
21. S. Späthe, F. Greulich, S. Apel, Context information management in a microservice based measurement and processing infrastructure, in *Innovations for Community Services: 22nd International Conference, I4CS 2022, Delft, The Netherlands, June 13–15, 2022, Proceedings* (Springer, New York, 2022), pp. 116–135
22. K.S. Swarnalatha, A. Mallya, G. Mukund, R.U. Bharadwaj, *Solving Problems of Large Codebases: Uber's Approach Using Microservice Architecture* (2022), pp. 653–662
23. A.F. da Silva Veloso, J.V.R. Júnior, R. de Andrade Lira Rabelo, J. Dela-flora Silveira, HyDSMaaS: a hybrid communication infrastructure with LoRaWAN and LoraMesh for the demand side management as a service. Future Internet **13**(11), 271 (2021)
24. M. Wittig, A. Wittig, *Amazon Web Services in Action* (2018)

Chapter 3
Heuristics for the Problem of Consolidating Orders into Vehicle Shipments with Compatible Categories and Freight Based on the Direct Distances to the Farthest Customers

Renan Sallai Iwayama and Claudio B. Cunha

Abstract In this chapter we address the problem of grouping orders into shipments for delivery that considers compatibility constraints as well as vehicle freight that is based on the direct distances to the farthest customers (PGOCCDD). This problem can be seen as a variant of the vehicle routing problem (VRP) that arises in the context of the last mile distribution to small stores and nanostores in developing countries, particularly in Latin America. Contrarily to the traditional VRP, in which the total distance traveled by vehicles is minimized, in our case the route costs depend solely on the direct distance to the delivery points that are located farthest with respect to the depot where vehicles originate. In addition, items to be delivered belong to categories of products that may not be compatible. To solve this problem, we propose a mathematical formulation to obtain exact solutions as well as a multi-start perturbation tabu (MSPT) neighborhood search heuristic that is composed of the greedy randomized adaptive search procedure (GRASP), tabu search (TS), and "iterated local search" (ILS). Our computational experiments show that the MSPT is competitive as it allowed us to obtain superior solutions in very short running times when compared to the exact method running for up to 5 hours in a high-performance computer.

R. S. Iwayama
Graduate Program in Logistics Systems Engineering, University of São Paulo, São Paulo, SP, Brazil

C. B. Cunha (✉)
Graduate Program in Logistics Systems Engineering, University of São Paulo, São Paulo, SP, Brazil

Universidade de São Paulo, Departamento de Eng de Transportes - Escola Politécnica, Ed Paula Souza (Prédio da Eng Civil) Cidade Universitária, São Paulo, Brazil
e-mail: cbcunha@usp.br

C. A. S. Oliveira, M. P. Pardalos (eds.), *Handbook of Artificial Intelligence and Data Sciences for Routing Problems*, Springer Optimization and Its Applications 219, https://doi.org/10.1007/978-3-031-78262-6_3

Keywords Heuristic · Combinatorial optimization · Routing · Urban transport · Goods distribution

3.1 Introduction

This chapter addresses a delivery problem that arises in the context of last mile distribution in developing countries, particularly for small stores in Latin America where millions of small, family-operated very small stores—henceforth addressed as nanostores—are the main source of products for daily needs from small in many neighborhoods of large cities, as well as at tiny towns and villages in remote rural areas [5]. Distributing these products, such as water, soap, and cooking oil, to nanostores is a challenging task, mainly because they have limited space and cash availability as well. Thus, shippers rely on wholesalers to ensure on-shelf availability of their products through consolidation that allows the delivery of modest amounts of a limited range of diverse items produced by different manufacturers.

Deliveries are usually performed by small- to mid-sized carriers which, in turn, oftentimes rely on hiring autonomous truck drivers that own and operate their sole vehicles, as the freight values they are willing to accept tend to be smaller due to a fierce competition as unemployment rates are high. As a result, freight delivery costs to be paid are usually calculated in a simplified and straightforward way: instead of taking into account the total traveled distance of each route, the due value is based solely on the direct distance between the depot and the delivery point that is farthest from it among those assigned to the vehicle. It should also be noted that, in general, given the location of the retail outlets to be serviced, the distances of line-haul legs of the delivery routes (i.e., distance between the depot and the delivery zone) are significantly greater than those within the delivery area. Therefore, the fixed costs of the delivery vehicles prevail over the variable ones, and determining the best route in terms of sequencing the points to be visited by a vehicle is usually left to its driver, once the due payment does not depend on the actual mileage performed by the vehicle.

This informal practice, in a sector that relies mostly on third parties, and sometimes lacks even fundamental logistics management tools such as a TMS (transportation management systems), allows simplifying the required paperwork related to the processes of validating and auditing the payment of the invoices submitted by carriers and independent truck owners. However, some practical constraints must be respected to avoid assigning a set of dispersed, not well-clustered deliveries that will require a total distance to be traveled that considerably increases the direct distance between the depot and the farthest retail outlet, since drivers may tend to not accept such unfavorable set of deliveries. This assessment is usually performed visually by the assigned driver, based on the maximum threshold azimuth difference, i.e., the angle between two distinct lines, each one linking a delivery point and the depot, which can be determined taking into consideration the polar coordinates of the delivery points, having the depot as the center.

In addition, in the traditional VRP there is no consideration of the product characteristics, and this does not interfere with the minimization objective of the problem. However, it is possible to generalize this approach and characterize similar products by category. This is an important consideration in cases in which it is necessary to segregate the transport of products with incompatible categories due to good practices or by the imposition of regulatory agencies. For example, hygiene and cleaning products must not be transported with food products.

The mathematical formulation we propose for this specific problem that we address, called the problem of grouping orders into shipments with compatible categories and freight based on the direct distances to the farthest customers (PGOCCDD), could be derived solely from the vehicle routing problem (VRP). However, given that the freight payment due to the carrier (or logistic operator) on a route does not consider the total distance to be traveled on the route, we also take into account the capacitated clustering problem (CCP), as it does not consider the sequence of delivery stops assigned to each vehicle. The PGOCCDD is NP-hard since it can be can be modeled as CCP, which is also NP-hard [6] if all products belong to a single category or no incompatibility is present.

To solve the problem, we propose a multi-start perturbation tabu neighborhood search (MSPT) heuristic. Eight different procedures to generate initial solutions are evaluated, and some simple local search heuristics are designed to better exploit and explore the solution space.

The remaining of this chapter is organized as follows: a literature review on related problems is presented in Sect. 3.2; the characterization of the problem and its mathematical formulation is given in Sect. 3.3 the heuristics for solving the problem are described in Sect. 3.4; Sect. 3.5 contains the computational experiments, results, and discussions; and finally, concluding remarks and future research directions are presented in Sect. 3.6.

3.2 Literature Review

To the best of our knowledge, neither the PGOCCDD nor other similar problems have been addressed in the literature. Thus, we focus our literature on related problems as the vehicle routing problem (VRP), the capacitated clustering problem (CCP), as well as the bin packing problem (BPP). On the other hand, groupings of orders define delivery regions, and, in this way, the methods used to solve the CCP can be useful. Similarly, a good solution should also have as few vehicles as possible with high occupancy since the freight is directly proportional to the number of vehicles and so it is interesting to explore the solution methods for BPP.

Several methods for VRP resolution can be found in the literature. The reader is referred to the works of Laporte et al. [14], Toth and Vigo [22], and Golden et al. [8] for good general reviews.

Vidal et al. [23] list different solution attributes that may arise in a variety of "multi-attribute vehicle routing problems" (MAVRPs). However, in our chapter, we address two distinct attributes that have not yet been found in the VRP literature: the consideration of freight by the maximum direct distance and a threshold angle on a route.

Applications of the capacitated clustering problem (CCP) can be found in the context of practical situations such as assigning customer orders into vehicle shipments [3, 13], topological design of computer communication networks [18], definition of set of customers to be supplied by a factory [17], or definition of territories for vendors [15]. As a working reference of clustering models and their applications, it is suggested to consult, for example, the works of Everitt et al. [2] and Kaufman and Rousseeuw [11].

Bin packing problems (BPPs) occur frequently in distribution and production processes in which it is necessary to group items into bins [4]. Quiroz-Castellanos et al. [19] proposed a heuristic named grouping genetic algorithm with controlled gene transmission (GGA-CGT) for the BPP that showed to be competitive with the state-of-the-art heuristics for very difficult instances.

Item-by-item conflicts were introduced in the BPP by Jansen and Ohring [10], but in the context of a scheduling problem in which the objects represent tasks and the boxes would be the task-processing machines. It was modeled considering a conflict graph where edges denote items that cannot be packaged together. This representation of the problem was also adopted by Gendreau and Laporte [7], Muritiba et al. [16], Jansen [9], Sadykov and Vanderbeck [20], and Khanafer et al. [12], all of which explicitly deal with the variant known as the bin packing problem with conflicts (BPC). Santos et al. [21] address the bin packing problem with compatible categories (BPCC), a distinct variant of the BPC that is directly to the PGOCCDD, as it also arises in the context of last mile distribution. It also considers the concept of incompatible categories of items to be delivered (i.e., types of items that cannot be transported together, such as food and cleaning products), as opposed to an item-by-item incompatibility in the BPC. Works related to these BPP may provide good insights to solve the PGOCCDD as their objective is to minimize the number of used bins, which can be seen as equivalent to the number of routes.

3.3 Problem Statement and Mathematical Formulation

As mentioned above, the PGOCCDD arises in the context of the last mile distribution to nanostores in urban centers. Similarly to Santos et al. [21], the orders placed by the stores have a large range of manufactures that may belong to distinct categories (sets of products with similar physical characteristics and packaging). We use the same categories (Table 3.1) and compatibility matrix (Table 3.2) as in Santos et al. [21].

The compatibility between product categories is an aspect that affects how deliveries are scheduled, since some distinct types of products cannot be transported together in the same vehicle. Such sort of restriction may be imposed either by codes of good practice or by regulatory institutions which, for example, would not allow the transport of food products together with cleaning materials.

Table 3.1 Relation of products by category

Category	Products
C1	Hygiene, cleaning, perfumery, and disposable goods
C2	Basic grocery (Oils, olive oil, pastas, chocolate, grains, farinaceous, coffee, rice, beans, and sugar)
C3	Cigarette and other tobacco goods
C4	Liquid grocery (alcoholic and non-alcoholic beverages)
C5	Eggs and FVG (fruits, vegetables, and greens)
C6	Complementary grocery (snacks, canned food, bonbonniere, biscuits, toast, cereals, and sweets)

Table 3.2 Matrix of compatibilities between product categories

	C_1	C_2	C_3	C_4	C_5	C_6
C_1	1	0	1	0	0	0
C_2	0	1	0	1	1	1
C_3	1	0	1	1	0	0
C_4	0	1	1	1	1	1
C_5	0	1	0	1	1	0
C_6	0	1	0	1	0	1

The typical problem we aim to solve comprises routing the deliveries for a single-day planning horizon. We assume that a sufficiently large fleet of capacitated delivery vehicles is available. An order to a delivery point (typically a retail store) comprises all items of a given category that should be delivered to the store; no split deliveries are allowed, i.e., all items of the same category must be delivered by the same vehicle. It should be noted, however, that a customer may eventually be serviced by more than one vehicle if it has ordered products of conflicting categories. Similarly, orders assigned to a vehicle must respect compatibility constraints. Every route starts and ends at the depot where the vehicles are loaded. No delivery time windows are present. All customers serviced by a vehicle must be within a maximum threshold angle measured having the depot as origin; as mentioned above, this practical constraint aims to avoid a large transversal or total distance. Lastly, the 2D or 3D packing of items in the vehicle is not considered.

3.3.1 Mathematical Formulation

The PGOCCDD can be formally defined as follows. Let $N = \{0, 1, \ldots, n\}$ be the set of $n + 1$ nodes and $A = \{(i, j) : i, j \in V, i \neq j\}$ be the set of edges. Let $V = \{1, 2, \ldots, NV\}$ be the set of available vehicles and $K = \{1, 2, \ldots, nk\}$ be the set of product categories. Node 0 represents the depot from where vehicles of a homogeneous fleet with capacity Q depart and return. Each of the other nodes $i \in N' = N \setminus \{0\}$ corresponds to an order, defined by a customer point and non-negative demand q_{ik} of a given product of category $k \in K$; in other words, $i \in N'$ represents the total amount of products of a given category $k \in K$ that needs to

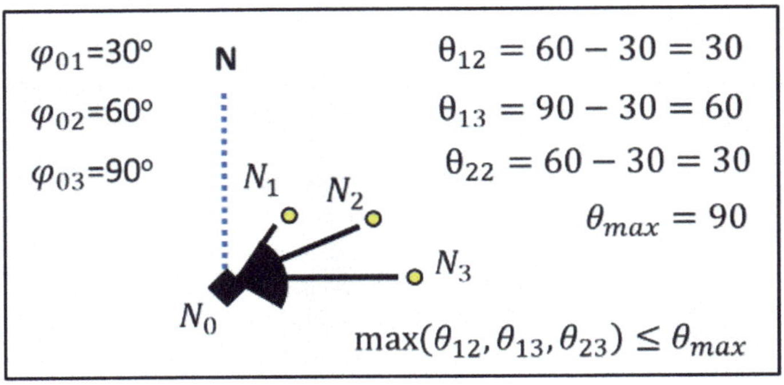

Fig. 3.1 Example of route angle and threshold angle

be delivered to a given customer location (a store). We also assume that $q_0^k = 0$, $\forall k \in K$. For each node $i \in N'$ its azimuth (i.e., its angular distance having the depot as origin) ϕ_i is known.

Each edge $(i, j) \in A$ has a non-negative cost c_{ij} that represents the distance between the locations of nodes i and j. Let $N'_k \subset N'$ be the subsets containing the orders $i \in N'$ such that $cat_i = c, \forall i \in N'_k$. The pair-wise compatibility between the categories is given by the binary matrix M'_{kk} where 1 denoted that two categories are compatible and zero otherwise. The angle formed by any two delivery points assigned to a vehicle is limited by ϕ_{max}; we also assume that all angles are in the range [0–180] degrees.

Figure 3.1 shows an example comprising a route with 3 delivery points whose azimuths are 30, 60, and 90 degrees. The maximum angle of the route is equal to 60 degrees (given by the maximum azimuth difference among all pairs of delivery points, i.e., θ_{13}), which is smaller than the threshold angle $\phi_m ax = 90°$. Finally, it is important to highlight that the number of available vehicles NV is sufficiently large and can be determined a priori as the same number of orders (i.e., $NV \geq k|N'|$), as we assume that $q_i k \leq Q, \forall i \in N' k \in K$. It should be noted that we use a heuristic to determine a reasonable upper bound on the number of vehicles to allow reducing the size of the formulation.

Decision variables:

- x_i^v, binary variable that equals 1 if order $i \in N'$ is assigned to vehicle $v \in V$ and zero otherwise
- f_k^v, binary variable that equals 1 if at least one order of category k is assigned to vehicle $v \in V$ and zero otherwise
- d^v, continuous variable that represent the maximum direct distance of vehicle $v \in V$

Objective function:

$$\min \sum_{v \in V} d^v \tag{3.1}$$

Subject to:

$$\sum_{v \in V} x_i^v = 1, \quad \forall i \in N' \tag{3.2}$$

$$\sum_{i \in N'} x_i^v q_i \leq Q, \quad \forall v \in V \tag{3.3}$$

$$f_k^v \geq x_i^v, \quad \forall v \in V, \forall k \in K, \forall i \in N_k' \tag{3.4}$$

$$f_k^v + f_{k'}^v \leq 1, \quad \forall v \in V, \forall k \in K, \forall k' \in K, M_{kk'} = 0, k < k' \tag{3.5}$$

$$x_i^v + x_j^v \leq 1, \quad \forall v \in V, \forall i \in N', \forall j \in N', \\ \min(|\phi_i - \phi_j|, 360 - |\phi_i - \phi_j|) > \phi_{\max} \tag{3.6}$$

$$d^v \geq x_i^v c_{0i}, \quad \forall v \in V, \forall i \in N' \tag{3.7}$$

$$d_v \geq d_{v'}, \quad \forall v \in V, \forall v' \in V, v < v' \tag{3.8}$$

$$x_i^v \in \{0, 1\}, \quad \forall i \in N, \forall v \in V \tag{3.9}$$

$$f_k^v \in \{0, 1\}, \quad \forall k \in K, \forall v \in V \tag{3.10}$$

$$d^v \in R, \quad \forall v \in V \tag{3.11}$$

The objective function (3.1) minimizes the total maximum direct distance of all used routes. Constraints (3.2) ensure that each request will be delivered and that this will be performed by a single vehicle. Constraints (3.3) ensure that the amount of cargo assigned to each vehicle does not exceed its capacity. Constraints (3.4) the variable f_k^v when at least one order of category k loaded in vehicle v. Together with constraints (3.5), they ensure that no vehicle carries products from conflicting categories. The maximum difference between the azimuths of delivery point pairs on a route is respected (i.e., is less than $\phi_{\max}$) and is given by constraints (3.6). Constraints (3.7) are of type min-max and allow determining the cost of each route (d^v) as the maximum direct distance on each route $v \in V$. Constraints (3.8) are not mandatory but were included with the aim of reducing the space of solutions by imposing that eventual empty vehicles are the last, thus preventing the symmetry in the enumeration tree to speed up the resolution of the problem. Finally, constraints (3.9), (3.10), and (3.11) are related to the domain of the decision variables.

3.4 Proposed Heuristic

To find good solutions in a timely manner for practical instances that may comprise hundreds of deliveries, we propose a multi-start iterated tabu search (MSPT) heuristic, which combines the following metaheuristics: GRASP [25], tabu search

Algorithm 1: MSPT

```
1 a ← 0;
2 b ← 0;
3 while a <= ω and b <= ω do
4     ChooseCH();
5     x ← x(hc);
6     if f(x) < f(x*) then
7         x* ← x;
8     end
9     if f(x) < f(x*) * ε then
10        a++;
11        j ← 0;
12        while j < τ do
13            if remainder(j/π) = 0 then
14                Shake();
15            end
16            Swap();
17            Replace();
18            if f(x) < f(x*) then
19                x* ← x;
20            end
21            j++
22        end
23    end
24    else
25        b++;
26    end
27 end
28 3Opt();
```

Fig. 3.2 Algorithm MSPT

(TS) [27], and iterated local search (ILS) [24]. The general scheme of the MSPT algorithm is presented in Fig. 3.2. It comprises a multi-start construction phase in which each new generated solution is then improved by a local search heuristic.

The construction phase consists of applying a constructive heuristic to obtain an initial solution that is chosen by ChooseCH (step 5); we consider a uniform probabilistic distribution among the eight proposed variants of classical constructive heuristics that are detailed in Sect. 4.1.

At the end of the construction phase an initial solution is created and the improvement phase is only initiated if the value of the objective function, calculated by the sum of the maximum direct distance of each route, is not greater than ε above the value of the best solution found (step 9). The construction phase is performed for ω restarts (step 3). The improvement phase is performed during τ iterations (step 12), in which procedures that diversify the use and exploration of the solution space are applied.

A three-opt procedure is then applied to improve the sequencing of delivery points in terms of minimizing the total distance traveled on each route as a secondary measure of the quality of the solution. It should be noted that the reduction of the total distance of each route is a secondary (desirable) objective of the problem that allows routes to be more attractive for the carriers, since it implies the reduction of the transportation variable costs.

3.4.1 Construction Phase

As the problem we address is, to the best of our knowledge, unprecedented in the literature, no specific constructive heuristic references were found in the literature for its solution. Thus, based on related methods for solving similar problems, eight different constructive heuristics are proposed, two of which are based on the vehicle routing problem (VRP) and six based on in the bin packing problem (BPP).

3.4.1.1 Constructive Heuristics for the VRP

Among the constructive heuristics for the VRP we decided to adapt the savings method proposed by Clarke and Wright. The classical savings method reduces the total distances traveled on each route by merging two routes into one. We performed savings calculation according to the expression (3.12):

$$S_{ij} = c_{i0} + c_{oj} - \lambda * c_{ij} + \gamma(q_i + q_j)/q^- + (\alpha * (2 * k - (gc_i + gc_j))/k) \quad (3.12)$$

- c_{i0}: distance between delivery point i and depot 0
- c_{0j}: distance between depot 0 and delivery point j
- c_{ij}: distance between delivery points i and j

In above expression (3.12), λ (shape factor) allows to adjust the emphasis that is placed on the distance between the delivery points i and j in relation to the direct distances between the depot from which the vehicles depart. We also defined two parcels in the calculation to help as a tie-breaking criterion for sorting the list of savings. One related to the demand that is adjusted by γ, named as demand factor, and one related to the compatibility between product categories that is adjusted by α, named as compatibility factor.

In expression (3.12), q_i and q_j are the demands related to delivery points i and j, respectively, and q^- is the average demand of all delivery points that is used to normalize the calculation. Also, gc_i and gc_j denote the product categories associated with delivery points i and j. The degree of incompatibility of a category is defined by counting the number of categories that are not compatible with itself and k is the total number of categories considered.

It is important to highlight that the classical expression of economies (Clarke and Wright, 1964) is obtained when $\lambda = 1$, $\gamma = 0$, and $\alpha = 0$. When $\lambda = 0$ the concept of saving may partially lose its meaning, but it might be that for the problem object of the present research, whose objective function considers only the direct distance from the depot to each of the delivery points, this consideration increases the chance of obtaining better results, since it favors the union of the points that are more distant from the depot and this leads to a lower cost than would be obtained if favored the union of points that are closer to each other, but not so far from the depot from where the vehicles depart. In addition, the maximum angle ensures that points that

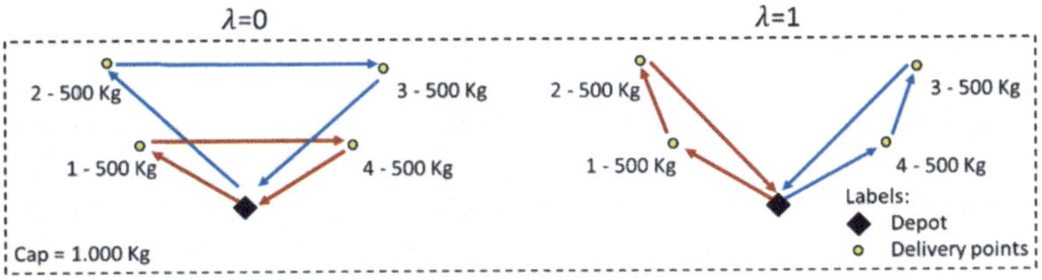

Fig. 3.3 Example of the influence of λ in the calculation of the savings

are transversely far apart are not merged into the same route. Figure 3.3 exemplifies what was described above, by means of an example in which it is possible to verify that considering $\lambda = 0$ provides a better result than for $\lambda = 1$, once in the first case routes are formed with delivery points that are in the same direction with respect to the depot since the distance between the delivery points is not considered.

Therefore, freight as a function of the most distant delivery point leads to less importance of the distance traveled between consecutive points on the route and emphasizes the better use of vehicles with the demand for delivery points that have similar direct distances to the depot.

Two versions of the savings method are evaluated: the sequential version named CW_SEQ, which strictly constructs one route at a time, and the parallel version named CW_PAR, which constructs routes simultaneously.

3.4.2 *Constructive Heuristics for the BPP*

A desirable feature of a good solution for the proposed problem is the lowest possible number of vehicles with high occupancy, since the cost of the solution is directly proportional to the number of vehicles.

Six constructive heuristics are proposed based on the methods proposed by Johnson [29] for the bin packing problem (BPP). In the "first fit" (FF) heuristic, each of the orders considered is sequentially included in the first feasible route; if no viable routes exist, a new route is created, and the procedure is repeated until all orders are assigned to a route. We also considered the "best fit" (BF) method, which includes a decreasing order in terms of occupation of the routes for each new order that is included in the solution. In addition, for both FF and BF heuristics, the sorting of the order list was considered at two levels, the first being the degree of incompatibility of the order product category (i.e., number of categories with which there is incompatibility) and second by the demand, distance to the depot, or azimuth of the order delivery point. FF-A and BF-A are similar to the sweep heuristic proposed by Gillett and Miller [26] for the VRP.

Different types of sorting seek to prioritize the orders that are more difficult to assign to existing routes. Therefore, it is considered a better option to first assign less flexible orders, as these would be more difficult to assign to existing routes

Table 3.3 Heuristics for BPP

		Order sorting	
		1st Level	2nd Level
Acronym	Route sorting	Ascending sorting	Descending sorting
FF-D	No sorting	Degree of compatibility	Demand
FF-DD	No sorting	Degree of compatibility	Distance to depot
FF-A	No sorting	Degree of compatibility	Azimuth
BF-D	Descending occupation	Degree of compatibility	Demand
BF-DD	Descending occupation	Degree of compatibility	Distance to depot
BF-A	Descending occupation	Degree of compatibility	Azimuth

already with different categories in future stages. Table 3.3 summarizes the six proposed heuristics resulting from the combination of different types of route and order sorting.

3.4.3 *Constructive Heuristics with GRASP*

The eight variants of classical constructive heuristics make the choice of the elements of construction of a solution in a sequential manner. However, inspired by the GRASP metaheuristic proposed by Feo and Rezende [25], the choice of the element that will be used to construct the solution (e.g., savings in the case of heuristics for the VRP, orders in the case of heuristics for the BPP) is randomized between a restricted list of candidates (RLC) containing the first not yet selected elements and the complete list of the building elements of a solution.

Therefore, following this logic, with each restart a new solution is generated, and this allows the construction phase alone to be able to produce solutions that can surpass the result of the classical constructive heuristics. In addition, the generation of different initial solutions allows the improvement phase to start at different points in the solution space, and this helps in the process of searching for better solutions.

3.5 Improvement Phase

MSPT improves an initial solution using both procedures that have the purpose of converging the result to a local optimum, local search procedure with TS, and for diversifying the solution with perturbations, shaking procedure. The sequence of application of the procedures is indicated in steps 13 to 17 described in Fig. 3.2. At each iteration, if the incumbent solution is greater than the best solution found so far, then the best solution is updated (step 18). Furthermore, after the last restart of the MSPT (step 28) the three-opt procedure is applied which is intended to define the delivery point sequence which minimizes the total distance traveled in each route and consequently reduces the variable transportation costs.

3.5.1 Local Search with TS

We perform local search using the replace and swap operators, inspired by the methods used by Osman [31]. However, instead of testing all combinations of movements between orders for two routes, the proposed procedure has the solution neighborhood to be explored limited by a certain criterion inspired by the granular search method used by Toth and Vigo [32].

When applying the replace operator, we consider only the set of orders of the first route to be tested with the following criteria: greater distance from the depot and the two orders with the extreme angles of the route. Orders that meet these criteria allow a change in the value of the objective function or at least help to make the angle of the route more flexible so that new feasible combinations of orders can be found.

The chosen order is reallocated in another route just after the depot (first stop of the route). Thus, if the reallocated order is the last one in its previous route, the number of routes of the solution is reduced. It is important to note that the choice to reallocate the chosen order in the second route at its first position was made to simplify and speed up the procedure, since the sequencing of orders does not change the objective function.

However, when the swap operator is applied all orders in the second route are considered as candidates to swap with the chosen order.

The proposed procedure is also inspired by TS metaheuristic use by Glover [27], and the choice of movement to be performed for both operators is made according to the first movement that generates improvement in the objective function or the one that generates the least degradation if the orders involved are not in the tabu period. The only exception is when a movement that contains an order in the tabu period can improve the objective function, because, in this case, the movement is accepted, characterizing the so-called aspiration criterion.

In both operators, the chosen order to be replaced or swapped is given a tabu period whose value is the squared root of the total number of orders n divided by two. In order to prevent the reversal of the movement in an efficient manner, the attributes of the solution are stored, which have been modified in previous iterations. Therefore, the order number with the maximum direct distance to the depot on route v is associated with the chosen order to be replaced or swapped, as attribute of the solution. If the order being moved is the client farthest from the depot, then the second order number is defined as the attribute of the solution.

The proposed procedure is applied by τ iterations but can be interrupted prematurely if a maximum number of consecutive iterations is reached without improvement of the objective function. The number of iterations without improvement is given by multiplying τ with σ and can assume values between 0 and 1.

3.5.2 Shaking Procedure

The proposed shaking procedure is inspired by ILS metaheuristic used by Hamdi-Dhaoui et al. [28] and aims to escape from local optimum by destroying and rebuilding part of a solution. There is a 50% chance for the operator to remove one or two categories of η% of the solution routes. The value η is randomly defined between 1 and the total number of routes of the solution each time the operator is applied. After the solution is partially destroyed, the orders that have been eliminated are reinserted in the solution following the logic of the FF heuristic. The frequency of use of this procedure is calculated by multiplying σ with π and can assume values between 0 and 1.

3.5.3 Three-Opt Procedure

The three-opt procedure is a special case of the $\lambda - opt$ procedure proposed by Lin [30] and consists of exchanging three arcs (or edges) in a route, reconnecting the delivery points in all other possible ways and evaluating each movement to find the best one and modify the solution accordingly. This process is then repeated for a different set of three arcs until there is no further possibility of reducing the distance traveled.

This procedure is used to find a good sequencing of deliveries that minimizes the distance on each route. It is applied only at the end, at step 23 in Fig. 3.2, since obtaining order sequencing during the improvement phase would bring an additional processing cost without bringing benefits to improve the objective function.

3.6 Computational Experiments

The proposed heuristics are implemented in C++ language using Microsoft Visual Studio 2017. A computer with an Intel® Core™ i7-4500U CPU @ 1.80 GHz processor and 8 Gbytes of RAM was used in the experiments.

We also performed computational experiments to solve the exact method using the high-performance computational resources of the Advanced Scientific Computation Laboratory (LCCA) of the University of São Paulo. More specifically, we used the CPLEX optimization software installed in the cluster Eagle that has 64 physical servers; each server has an Intel® Xeon® E7-2870 @ 2.40 GHz processor with 10 cores, 20 threads, and 512 GB of RAM. The processing time limit for each instance was set to 5 hours.

3.6.1 Description of Test Instances and Experiments

We selected 28 instances of routing problems from sets A and B proposed by Augerat [1] that are available at the CVRPLIB library (http://vrp.atd-lab.inf.puc-rio.br). These instances were created specifically for the vehicle routing problem (VRP), with a homogeneous fleet of capacitated vehicles. They have 31 to 78 clients and are named according to the default P_n N_{kK}, in which the first letter is the name of the set of instances, N is the number of nodes, and K indicates the number of routes to the best-known solution (BKS) by Augerat [1]. Set A has instances where the location of the warehouse, customers, and demands were randomly generated according to a uniform distribution, whereas the instances of set B are characterized by containing clustered client locations and so are more like practical cases of VRP. For all instances of sets A and B the capacity is 100 units and the demands are between 1 and 30 units.

All calculated distances and angles have been rounded to the nearest integer. To adapt the selected instances to the proposed problem, a division of the original demand into the six categories defined in a random fashion was made, so that the original demands were evenly distributed among the six categories. That is, an instance of 31 clients can have up to 186 orders and an instance of 78 clients can have up to 468 different orders.

3.6.2 Parameter Adjustment

Parameter selection influences the quality of the computational results. Therefore, we ran a full factorial experiment with three sets of parameters depicted in Table 3.4 (savings calculation—heuristics for the VRP, improvement phase, and multiple restarts), which correspond to the key parameters of the heuristics proposed in this research.

For this purpose, there are three instances of each of sets A and B, the smallest instance, a medium-sized instance, and the largest instance. The selected instances are the following: A-n32-k5, A-n39-k6, A-n48-k7, B-n31-k5, B-n50-k8, and B-n78-k10.

The results of these parameters adjustment indicate that the selected values on the last column of Table 3.4 seemed to provide best results.

As mentioned above, assuming $\lambda = 0$ suppresses a savings calculation, it is still a way of sorting the list of economies to build routes efficiently for the proposed problem. The use of the demand factor did not generate better results and, therefore, was not considered. On the other hand, the use of the compatibility factor made it possible to obtain better results for the proposed problem, once it considers the compatibility between categories of products to solve the allocation problem.

Table 3.4 Parameter adjustment

Type	Parameters	Symbol	Tested values	Selected
Savings calc.	Shape factor	λ	0 / 0.5 / 1 / 1.5 / 2	0
	Demand factor	γ	0 / 1	0
	Compatibility factor	α	0 / 1	1
Improv. phase	Number of iterations	τ	500 / 1000 / 1500	1500
	Number of iterations without improv.	σ	0.25 / 0.5 / 0.75	0.25
	Shake frequency	π	0.05 / 0.1 / 0.2	0.1
Multiple restarts	Number of restarts	ω	50 / 100 / 200	200
	Maximum dev. from the incumbent sol.	ε	1.1 / 1.2 / 1.3	1.3
	Size of restricted list of candidates	μ	3 / 6 / 9 /12	6

3.7 Results

The results of the computational experiments are divided into three subsections: instances with six categories, one category (i.e., it is assumed that all clients have only one order from the same category), and a third subsection that presents a comparative analysis between the optimal results of the original instances proposed by Augerat [1] with the objective of minimizing the total distance traveled and the results of the proposed heuristic method with the objective of minimizing the sum of the maximum direct distance on each route.

3.7.1 Instances with Six Categories

Tables 3.5 and 3.6 depict the results of each proposed constructive heuristic and the respective percentage difference with respect to the best solution obtained in each instance. The underlined values in bold highlight the heuristic that reached the lowest of the values among the others. Table 3.7 summarizes the results and it is possible to identify that the CW-PAR and BF-A heuristics ranked first with 28 best results. In addition, of the total best results, CW-PAR ranked first alone in 16 instances and CW-SEQ in 10 instances.

Therefore, these results indicate that it makes sense to adopt a multiple restart strategy in which a random choice occurs that leads to a diversification of the constructive heuristic to be used for the method to be robust in a wide variety of instances.

The processing time is not shown, since it is low and there is not much difference among the eight constructive heuristics. The processing time is less than one second most of the time and reaches almost three seconds in the worst case (Instance n66-k9-90 of set B).

The results of the exact method obtained with the CPLEX are given in Tables 3.8 and 3.9 in conjunction with the results of the heuristic method.

Table 3.5 Constructive heuristic results: six categories (set A)

Inst Set A	CW-SEQ	CW-PAR	FF-P	FF-D	FF-A	BF-P	BF-D	BF-A
n32-k5-90	256—8%	**236—0%**	289—22%	249—6%	259—10%	316—34%	255—8%	279—18%
n32-k5-60	292—4%	292—4%	317—13%	305—9%	**281—0%**	314—12%	304—8%	285—1%
n32-k5-45	322—13%	322—13%	317—12%	302—6%	**284—0%**	317—12%	302—6%	318—12%
n33-k5-90	238—10%	**217—0%**	276—27%	238—10%	221—2%	276—27%	238—10%	**217—0%**
n33-k5-60	302—35%	302—35%	314—41%	302—35%	**223—0%**	314—41%	302—35%	224—0%
n33-k5-45	311—1%	311—1%	309—1%	311—1%	311—1%	309—1%	311—1%	**307—0%**
n33-k6-90	**217—0%**	**217—0%**	261—20%	**217—0%**	221—2%	252—16%	**217—0%**	**217—0%**
n33-k6-60	310—39%	310—39%	280—26%	310—39%	**223—0%**	312—40%	310—39%	**223—0%**
n33-k6-45	332—6%	**313—0%**	342—9%	332—6%	**313—0%**	342—9%	332—6%	**313—0%**
n34-k5-90	**217—0%**	**217—0%**	291—34%	**217—0%**	221—2%	282—30%	**217—0%**	**217—0%**
n34-k5-60	310—39%	310—39%	310—39%	310—39%	**223—0%**	310—39%	310—39%	**223—0%**
n34-k5-45	332—6%	332—6%	**313—0%**	332—6%	**313—0%**	**313—0%**	332—6%	**313—0%**
n36-k5-90	**217—0%**	**217—0%**	291—34%	**217—0%**	221—2%	282—30%	**217—0%**	**217—0%**
n36-k5-60	310—39%	310—39%	310—39%	310—39%	**223—0%**	342—53%	310—39%	**223—0%**
n36-k5-45	332—6%	**313—0%**	342—9%	332—6%	**313—0%**	342—9%	332—6%	**313—0%**
n37-k5-90	384—4%	**370—0%**	441—19%	427—15%	445—20%	479—29%	432—17%	434—17%
n37-k5-60	493—6%	**467—0%**	522—12%	488—4%	535—15%	554—19%	468—0%	529—13%
n37-k5-45	673—14%	673—14%	609—3%	676—15%	**590—0%**	609—3%	700—19%	**590—0%**
n37-k6-90	317—0%	**316—0%**	401—27%	371—17%	342—8%	416—32%	371—17%	368—16%
n37-k6-60	334—2%	**326—0%**	374—15%	362—11%	346—6%	413—27%	393—21%	364—12%
n37-k6-45	376—3%	**366—0%**	389—6%	371—1%	383—5%	386—5%	378—3%	383—5%
n38-k5-90	460—14%	460—14%	445—10%	470—17%	**403—0%**	463—15%	452—12%	**403—0%**
n38-k5-60	590—7%	656—19%	617—12%	597—8%	557—1%	623—13%	590—7%	**551—0%**
n38-k5-45	723—5%	723—5%	773—13%	705—3%	**686—0%**	761—11%	688—0%	691—1%

n39-k5-90	495—13%	504—15%	475—8%	505—15%	**438—0%**	493—13%	481—10%	**438—0%**
n39-k5-60	671—8%	693—11%	670—8%	662—6%	628—1%	698—12%	688—11%	**622—0%**
v n39-k5-45	774—7%	754—4%	832—15%	756—5%	740—2%	818—13%	**723—0%**	741—2%
n39-k6-90	504—19%	546—29%	**423—0%**	529—25%	452—7%	**423—0%**	496—17%	448—6%
n39-k6-60	611—3%	**596—0%**	678—14%	599—1%	600—1%	670—12%	**596—0%**	599—1%
n39-k6-45	783—8%	783—8%	788—9%	787—9%	**724—0%**	754—4%	831—15%	732—1%
n45-k6-90	502—19%	546—29%	**423—0%**	523—24%	448—6%	**423—0%**	496—17%	448—6%
n45-k6-60	603—1%	**596—0%**	678—14%	**596—0%**	600—1%	670—12%	**596—0%**	599—1%
n45-k6-45	783—8%	783—8%	788—9%	787—9%	**724—0%**	754—4%	831—15%	732—1%
n45-k7-90	600—8%	600—8%	710—28%	643—16%	**554—0%**	718—30%	603—9%	557—1%
n45-k7-60	852—12%	852—12%	814—7%	852—12%	**762—0%**	814—7%	852—12%	**762—0%**
n45-k7-45	879—13%	879—13%	891—14%	879—13%	**780—0%**	891—14%	879—13%	**780—0%**
n46-k7-90	565—8%	**521—0%**	623—20%	575—10%	574—10%	598—15%	529—2%	574—10%
n46-k7-60	789—18%	831—24%	774—16%	789—18%	**669—0%**	765—14%	822—23%	**669—0%**
n46-k7-45	857—0%	872—2%	1014—19%	**853—0%**	954—12%	1006—18%	867—2%	948—11%
n48-k7-90	647—7%	645—7%	660—9%	633—5%	712—18%	**603—0%**	645—7%	681—13%
n48-k7-60	791—5%	791—5%	820—9%	791—5%	**754—0%**	797—6%	791—5%	**754—0%**
n48-k7-45	933—7%	933—7%	**874—0%**	933—7%	885—1%	**874—0%**	949—9%	885—1%
Total	21287—6%	21301—6%	22068—10%	21443—6%	**20135—0%**	22096—10%	21436—6%	20171—0%

Table 3.6 Constructive heuristic results: six categories (set B)

Inst Set B	CW-SEQ	CW-PAR	FF-P	FF-D	FF-A	BF-P	BF-D	BF-A
n31-k5-90	586—9%	**536—0%**	647—21%	556—4%	713—33%	621—16%	596—11%	687—28%
n31-k5-60	837—8%	890—15%	869—12%	837—8%	778—0%	851—10%	837—8%	**777—0%**
n31-k5-45	**921—0%**	1017—10%	997—8%	**921—0%**	933—1%	997—8%	**921—0%**	933—1%
n34-k5-90	723—6%	833—23%	**680—0%**	732—8%	813—20%	742—9%	710—4%	705—4%
n34-k5-60	773—2%	849—12%	1053—39%	**755—0%**	928—23%	1019—35%	799—6%	899—19%
n34-k5-45	1063—26%	1100—30%	927—10%	1063—26%	857—1%	927—10%	1076—27%	**845—0%**
n35-k5-90	612—7%	718—25%	758—32%	612—7%	**573—0%**	758—32%	633—10%	**573—0%**
n35-k5-60	822—2%	1029—28%	840—4%	816—1%	**807—0%**	840—4%	820—2%	**807—0%**
n35-k5-45	**930—0%**	**930—0%**	1171—26%	**930—0%**	1008—8%	1171—26%	975—5%	1008—8%
n38-k6-90	**734—0%**	736—0%	854—16%	819—12%	939—28%	860—17%	796—8%	900—23%
n38-k6-60	**819—0%**	**819—0%**	1007—23%	841—3%	829—1%	989—21%	832—2%	824—1%
n38-k6-45	**945—0%**	**945—0%**	1169—24%	**945—0%**	1077—14%	1150—22%	958—1%	1090—15%
n39-k5-90	577—5%	615—12%	630—15%	**547—0%**	699—28%	606—11%	575—5%	699—28%
n39-k5-60	**783—0%**	**783—0%**	883—13%	**783—0%**	915—17%	883—13%	**783—0%**	915—17%
n39-k5-45	1011—3%	1011—3%	1003—2%	1011—3%	**984—0%**	1003—2%	1011—3%	**984—0%**
n41-k6-90	721—9%	748—13%	819—24%	752—14%	668—1%	875—32%	741—12%	**661—0%**
n41-k6-60	827—3%	987—23%	874—9%	807—1%	862—8%	927—16%	**800—0%**	862—8%
n41-k6-45	965—4%	1087—17%	1047—13%	970—5%	972—5%	1153—24%	1013—9%	**928—0%**
n43-k6-90	617—2%	608—0%	819—35%	**607—0%**	843—39%	804—32%	608—0%	839—38%
n43-k6-60	894—3%	932—7%	1027—18%	899—3%	903—4%	1016—17%	**870—0%**	903—4%
n43-k6-45	1014—3%	1208—23%	1267—29%	1029—5%	1029—5%	1267—29%	**984—0%**	1029—5%
n44-k7-90	729—4%	747—7%	813—16%	741—6%	**701—0%**	779—11%	791—13%	742—6%
n44-k7-60	856—4%	875—6%	897—9%	**822—0%**	894—9%	874—6%	**822—0%**	888—8%

n44-k7-45	991—18%	991—18%	1063—27%	991—18%	**840—0%**	1081—29%	1008—20%	**840—0%**
n45-k5-90	2110—0%	**2102—0%**	2930—39%	2531—20%	2727—30%	2834—35%	2596—24%	2703—29%
n45-k5-60	**2214—0%**	2219—0%	2733—23%	2642—19%	2768—25%	2808—27%	2618—18%	2738—24%
n45-k5-45	2358—1%	**2342—0%**	2936—25%	2785—19%	2809—20%	2924—25%	2737—17%	2851—22%
n45-k6-90	**2390—0%**	2397—0%	3404—42%	3099—30%	3472—45%	3448—44%	3318—39%	3440—44%
n45-k6-60	**2482—0%**	2492—0%	3487—40%	3190—29%	3465—40%	3573—44%	3169—28%	3419—38%
n45-k6-45	**2506—0%**	2539—1%	3430—37%	3172—27%	3388—35%	3435—37%	3265—30%	3358—34%
n50-k7-90	1909—0%	**1908—0%**	2597—36%	2333—22%	2538—33%	2593—36%	2372—24%	2534—33%
n50-k7-60	**1972—0%**	2034—3%	2593—31%	2417—23%	2486—26%	2586—31%	2354—19%	2500—27%
n50-k7-45	**2103—0%**	2124—1%	2607—24%	2438—16%	2593—23%	2598—24%	2472—18%	2563—22%
n50-k8-90	2402—1%	**2374—0%**	3132—32%	2801—18%	3133—32%	3164—33%	2884—21%	3212—35%
n50-k8-60	**2424—0%**	2452—1%	3152—30%	2932—21%	3120—29%	3208—32%	2972—23%	3115—29%
n50-k8-45	**2544—0%**	2575—1%	3218—26%	2963—16%	3082—21%	3215—26%	3060—20%	3055—20%
n66-k9-90	**1487—0%**	1490—0%	1953—31%	1639—10%	1736—17%	1974—33%	1659—12%	1715—15%
n66-k9-60	1487—2%	**1452—0%**	1816—25%	1596—10%	1772—22%	1833—26%	1584—9%	1676—15%
n66-k9-45	**1520—0%**	1562—3%	1732—14%	1654—9%	1745—15%	1731—14%	1676—10%	1763—16%
n78-k10-90	1756—0%	**1755—0%**	1905—9%	1863—6%	1822—4%	1948—11%	1919—9%	1773—1%
n78-k10-60	1735—0%	**1731—0%**	1952—13%	1843—6%	1865—8%	1903—10%	1845—7%	1839—6%
n78-k10-45	1735—0%	**1730—0%**	1815—5%	1771—2%	1816—5%	1825—5%	1768—2%	1816—5%
Total	**56884—0%**	58272—2%	69506—22%	63455—12%	66902—18%	69793—23%	64227—13%	66408—17%

Table 3.7 Summary of constructive heuristics: six categories

Set	CW-SEQ	CW-PAR	FF-P	FF-D	FF-A	BF-P	BF-D	BF-A
A	3	15	4	5	20	5	6	20
B	16	13	1	8	5	0	6	8
Total	19	28	5	13	25	5	12	28

In each of these tables, the columns below the optimal header contain the results for exact method. The "BKS" column indicates the value of the objective function of best solution obtained (highlighted in bold when it is the optimal value). The column "% GAP LB" indicates the percentage difference between the "lower bound" and the "BKS." The time column indicates the processing time that is at the limit of 18,000 seconds (5 hours). Then, the next columns indicate results of the heuristic method. The "CF" column indicates the percentage of the value obtained using the best constructive heuristic for the construction phase with respect to "BKS."

The "CF + IF" column indicates the percentage of the value obtained when using the best constructive heuristic for the construction phase and the improvement phase in relation to the "BKS." The columns below the header "MSPT" refer to the results of the MSPT heuristic. The "Min" column indicates the best MSPT result (highlighted in bold for values less than or equal to "BKS") using ten random seeds for random number generation. The "% BKS" column indicates the percentage of the minimum value obtained using MSPT with respect to "BKS." The "Average" column indicates the average of the results among the ten seeds used. Finally, the time column indicates the average processing time for each instance.

Table 3.10 presents a summary of the CPLEX results and it is possible to identify that it was possible to generate 7 optimal solutions (column Opt) and 62 feasible solutions (column FS). Among 84 instances, in 15 cases it was not possible to generate a feasible solution (column n/a) along the processing time of 5 h.

According to Table 3.11, the result of the best constructive heuristic is greater than or equal to the BKS in 66 instances (79% of the total) which demonstrates that the proposed heuristics have better or equal quality than the exact method. This value increases to 71 instances (85% of the total) when the improvement phase is applied and increases further to 76 instances (90% of the total) when the MSPT is used. Therefore, it is possible to affirm that the use of the improvement phase and the multi-start strategy improved the results. The instances n36-k5-90 of set A are the one with the highest "% BKS" (5.5%) and the only instance above 5%.

Therefore, as explained in the above analysis, it is possible to verify that due to the combinatorial complexity of the proposed problem, it is not possible to obtain a solution with quality assurance in terms of the objective function and processing time using the exact method. On the other hand, the proposed heuristics could generate feasible solutions in all tested instances.

Table 3.8 Exact and heuristic results for six categories (Augerat's subset A)

Inst	CPLEX			CP	CP+IP	MSPT			
Set A	BKS	%GAP LB	Time	%BKS		Min	%BKS	Average	Time
n32-k5-90	**534**	0%	236	3.7%	0.0%	**534**	0.0%	534.0	32
n32-k5-60	**744**	0%	9.091	2.4%	2.0%	**744**	0.0%	744.0	34
n32-k5-45	774	64%	18.000	0.8%	0.0%	**774**	0.0%	774.0	35
n33-k5-90	521	67%	18.000	0.0%	0.0%	**521**	0.0%	521.0	45
n33-k5-60	684	83%	18.000	−2.2%	−2.2%	**669**	−2.2%	669.0	47
n33-k5-45	1.324	94%	18.000	−35.6%	−35.7%	**851**	−35.7%	851.0	57
n33-k6-90	537	59%	18.000	12.3%	5.2%	**531**	−1.1%	533.0	45
n33-k6-60	772	80%	18.000	−2.3%	−3.5%	**745**	−3.5%	745.0	48
n33-k6-45	1.152	92%	18.000	−24.1%	−26.2%	**850**	−26.2%	850.0	56
n34-k5-90	**524**	0%	3.580	2.3%	2.3%	536	2.3%	536.0	40
n34-k5-60	792	80%	18.000	−1.9%	−9.2%	**719**	−9.2%	719.0	49
n34-k5-45	1.292	93%	18.000	−28.7%	−29.6%	**909**	−29.6%	909.0	57
n36-k5-90	**549**	0%	81	23.9%	5.5%	579	5.5%	579.0	43
n36-k5-60	673	67%	18.000	12.2%	−1.2%	**665**	−1.2%	665.0	44
n36-k5-45	996	85%	18.000	−15.2%	−15.2%	**842**	−15.5%	842.0	48
n37-k5-90	579	37%	18.000	−1.0%	−1.0%	**573**	−1.0%	573.0	42
n37-k5-60	870	82%	18.000	−7.2%	−8.3%	**798**	−8.3%	798.0	53
n37-k5-45	1.465	94%	18.000	−36.5%	−36.5%	**930**	−36.5%	930.0	62
n37-k6-90	597	13%	18.000	22.9%	2.2%	**597**	0.0%	597.9	24
n37-k6-60	800	81%	18.000	2.4%	−3.5%	**762**	−4.8%	762.0	53
n37-k6-45	1.669	92%	18.000	−43.4%	−50.4%	**828**	−50.4%	828.0	58
n38-k5-90	547	70%	18.000	0.0%	0.0%	**547**	0.0%	547.0	44
n38-k5-60	1.269	93%	18.000	−38.3%	−40.5%	**741**	−41.6%	741.0	57
n38-k5-45	1.488	94%	18.000	−33.9%	−36.7%	**942**	−36.7%	942.0	70
n39-k5-90	**577**	0%	12.515	14.6%	5.4%	**577**	0.0%	580.6	27
n39-k5-60	869	86%	18.000	−7.9%	−17.3%	**718**	−17.4%	718.0	50
n39-k5-45	1.204	90%	18.000	−22.9%	−24.7%	**894**	−25.7%	895.1	59
n39-k6-90	655	65%	18.000	−7.3%	−9.0%	**596**	−9.0%	596.0	51
n39-k6-60	1.287	92%	18.000	−32.4%	−36.8%	**814**	−36.8%	814.0	61
n39-k6-45	1.692	94%	18.000	−41.8%	−41.8%	**971**	−42.6%	971.0	66
n45-k6-90	3.644	97%	18.000	−42.3%	−42.3%	**2.090**	−42.6%	2.095.3	238
n45-k6-60	n/a	n/a	18.000	n/a	n/a	**2.182**	n/a	2.189.4	258
n45-k6-45	4.185	98%	18.000	−44.0%	−44.5%	**2.250**	−46.2%	2.267.7	250
n45-k7-90	3.248	97%	18.000	−26.4%	−26.4%	**2.385**	−26.6%	2.391.1	111
n45-k7-60	4.018	97%	18.000	−38.2%	−38.2%	**2.458**	−38.8%	2.482.4	110
n45-k7-45	n/a	n/a	18.000	n/a	n/a	**2.528**	n/a	2.530.6	177
n46-k7-90	2.996	97%	18.000	−36.3%	−36.3%	**1.890**	−36.9%	1.900.1	210
n46-k7-60	n/a	n/a	18.000	n/a	n/a	**1.982**	n/a	1.994.0	275
n46-k7-45	n/a	n/a	18.000	n/a	n/a	**2.038**	n/a	2.068.9	262
n48-k7-90	3.630	97%	18.000	−34.6%	−34.6%	**2.378**	−34.5%	2.385.9	306
n48-k7-60	n/a	n/a	18.000	n/a	n/a	**2.416**	n/a	2.422.4	317
n48-k7-45	n/a	n/a	18.000	n/a	n/a	**2.520**	n/a	2.529.4	294

Table 3.9 Exact and heuristic results: six categories (set B)

Inst	CPLEX			CP	CP+IP	MSPT			
Set B	BKS	%GAP LB	Time	%BKS		Min	%BKS	Average	Time
n31-k5-90	1.419	22%	18.000	4.8%	4.7%	1.451	2.3%	1.456.6	190
n31-k5-60	1.428	86%	18.000	1.7%	1.7%	1.455	1.9%	1.458.4	177
n31-k5-45	1.725	92%	18.000	−11.9%	−11.9%	**1.526**	−11.5%	1.530.0	173
n34-k5-90	1.649	89%	18.000	6.4%	1.6%	**1.647**	−0.1%	1.666.1	203
n34-k5-60	2.026	95%	18.000	−14.6%	−14.8%	**1.706**	−15.8%	1.715.4	177
n34-k5-45	2.017	96%	18.000	−14.2%	−14.2%	**1.709**	−15.3%	1.714.9	176
n35-k5-90	2.204	93%	18.000	0.4%	−0.3%	**2.126**	−3.5%	2.135.7	195
n35-k5-60	2.633	95%	18.000	−16.6%	−16.6%	**2.146**	−18.5%	2.174.1	168
n35-k5-45	n/a	n/a	18.000	n/a	n/a	**2.136**	n/a	2.149.9	149
n38-k6-90	1.726	94%	18.000	−6.2%	−6.2%	**1.607**	−6.9%	1.612.4	125
n38-k6-60	2.397	96%	18.000	−31.1%	−31.1%	**1.659**	−30.8%	1.670.9	194
n38-k6-45	2.385	90%	18.000	−29.7%	−29.7%	**1.644**	−31.1%	1.667.6	176
n39-k5-90	2.243	96%	18.000	−32.0%	−32.0%	**1.504**	−32.9%	1.509.6	64
n39-k5-60	2.441	96%	18.000	−35.3%	−35.6%	**1.515**	−37.9%	1.530.9	104
n39-k5-45	2.720	96%	18.000	−41.9%	−41.9%	**1.509**	−44.5%	1.530.3	107
n41-k6-90	2.727	97%	18.000	−36.9%	−36.9%	**1.682**	−38.3%	1.697.6	240
n41-k6-60	2.812	97%	18.000	−38.0%	−38.0%	**1.720**	−38.8%	1.736.4	222
n41-k6-45	n/a	n/a	18.000	n/a	n/a	**1.753**	n/a	1.768.7	209
n43-k6-90	2.363	97%	18.000	−31.9%	−31.9%	**1.568**	−33.6%	1.580.3	254
n43-k6-60	n/a	n/a	18.000	n/a	n/a	**1.596**	n/a	1.602.1	232
n43-k6-45	n/a	n/a	18.000	n/a	n/a	**1.622**	n/a	1.642.7	234
n44-k7-90	2.468	96%	18.000	−25.9%	−25.9%	**1.835**	−25.6%	1.842.9	110
n44-k7-60	2.856	97%	18.000	−36.1%	−36.1%	**1.833**	−35.8%	1.836.3	132
n44-k7-45	3.024	97%	18.000	−38.8%	−38.8%	**1.879**	−37.9%	1.886.9	204
n45-k5-90	3.076	97%	18.000	−46.3%	−46.3%	**1.622**	−47.3%	1.628.9	177
n45-k5-60	3.219	97%	18.000	−48.1%	−48.1%	**1.644**	−48.9%	1.661.9	239
n45-k5-45	n/a	n/a	18.000	n/a	n/a	**1.763**	n/a	1.778.9	239
n45-k6-90	2.356	97%	18.000	−27.9%	−27.9%	**1.663**	−29.4%	1.672.6	356
n45-k6-60	2.571	97%	18.000	−31.9%	−31.9%	**1.724**	−32.9%	1.729.0	343
n45-k6-45	2.795	98%	18.000	−34.9%	−34.9%	**1.788**	−36.0%	1.797.4	320
n50-k7-90	n/a	n/a	18.000	n/a	n/a	**1.771**	n/a	1.782.1	324
n50-k7-60	n/a	n/a	18.000	n/a	n/a	**1.851**	n/a	1.865.3	286
n50-k7-45	n/a	n/a	18.000	n/a	n/a	**1.850**	n/a	1.868.3	288
n50-k8-90	**624**	0%	3.286	10.9%	6.4%	642	2.9%	649.6	37
n50-k8-60	**726**	0%	1.861	16.3%	5.8%	746	2.8%	756.7	72
n50-k8-45	880	74%	18.000	7.3%	5.8%	893	1.5%	896.7	85
n66-k9-90	730	69%	18.000	17.9%	7.8%	759	4.0%	764.3	122
n66-k9-60	1.183	83%	18.000	−25.8%	−29.0%	**819**	−30.8%	827.3	109
n66-k9-45	1.621	93%	18.000	−41.2%	−43.5%	**906**	−44.1%	911.7	109
n78-k10-90	990	88%	18.000	−27.0%	−28.7%	**648**	−34.5%	666.6	104
n78-k10-60	1.350	92%	18.000	−40.1%	−44.8%	**742**	−45.0%	742.1	111
n78-k10-45	n/a	n/a	18.000	n/a	n/a	**823**	n/a	824.1	119

Table 3.10 Summary of exact and heuristic results: six categories

Set	Opt	FS	n/a	Total
A	5	31	6	42
B	2	31	9	42
Total	7	62	15	84

Table 3.11 Summary of heuristics %BKS: six categories

%BKS	CP	CP+IP	MSPT
≤ 0%	66	71	76
0%–1%	2	0	0
1%–5%	6	6	7
> 5%	10	7	1

3.7.2 Instances with One Category

To evaluate the quality of heuristics in relation to the problem without categories, a comparison is made considering the classic problem of vehicle routing that does not have this characteristic (the threshold angle attribute is still considered). That is, it is assumed that the original instances proposed by Augerat [1] have only one category.

The results of the exact method obtained with CPLEX are shown in Tables 3.12 and 3.13 along with the results using the MSPT.

In each of these tables, the columns below the "CPLEX" heading contain exact method results. The "BKS" column indicates the best CEPLEX solution generated (value in bold when optimal is found). The column "% GAP LB" indicates the percentage difference between the "lower bound" and the "BKS." The time column indicates the processing time that is at the limit of 18,000 seconds (5 hours). Then, the next columns indicate results of the heuristic method. The columns below the header "MSPT" refer to the results of the MSPT heuristic. The Min column indicates the best MSPT result using ten random seeds for random number generation. The "% BKS" column indicates the percentage of the minimum value obtained using MSPT with respect to "BKS." The "Average" column indicates the average of the result among the ten seeds used. Finally, the time column indicates the average processing time of each instance.

Table 3.14 presents a summary of the exact results, showing 60 optimal solutions and 24 feasible solutions. In no case the exact method failed to generate an unfeasible solution in the processing time of 5h. The results of set *A* presented 34 optimal solutions and the results of set *B* presented 26 optimal solutions. Therefore, we can state that set *B* has more difficult instances.

However, if compared to the problem considering six categories, it is possible to affirm that in both sets the problem is easier to solve because its dimension is reduced six times. When considering only 1 category, it was possible to generate a greater number of optimal solutions, and analyzing the result of the "% GAP LB" column it is possible to show that just in a few instances this value is high (n66-k9-90 and n78-k10-90 of set B).

Table 3.12 CPLEX and MSPT results: one category (set A)

Inst	CPLEX			MSPT			
Set A	BKS	%GAP LB	Time	Min	%BKS	Average	Time
n32-k5-90	**311**	0.0%	2	**311**	0.0%	311.0	5
n32-k5-60	**325**	0.0%	3	**325**	0.0%	325.4	4
n32-k5-45	**350**	0.0%	3	**350**	0.0%	350.0	4
n33-k5-90	**258**	0.0%	4	**258**	0.0%	258.0	4
n33-k5-60	**292**	0.0%	4	**292**	0.0%	292.0	4
n33-k5-45	**332**	0.0%	5	**332**	0.0%	332.0	4
n33-k6-90	**275**	0.0%	6	**275**	0.0%	275.0	5
n33-k6-60	**327**	0.0%	7	**327**	0.0%	327.0	5
n33-k6-45	**357**	0.0%	8	**357**	0.0%	357.0	4
n34-k5-90	**314**	0.0%	3	**314**	0.0%	314.0	6
n34-k5-60	**331**	0.0%	4	**331**	0.0%	331.0	5
n34-k5-45	**353**	0.0%	5	**353**	0.0%	356.0	5
n36-k5-90	**310**	0.0%	3	**310**	0.0%	310.0	5
n36-k5-60	**333**	0.0%	4	**333**	0.0%	333.0	5
n36-k5-45	**347**	0.0%	3	**347**	0.0%	347.0	5
n37-k5-90	**241**	0.0%	3	**241**	0.0%	241.0	4
n37-k5-60	**289**	0.0%	4	**289**	0.0%	289.0	4
n37-k5-45	**329**	0.0%	6	**329**	0.0%	329.0	4
n37-k6-90	372	0.3%	7	**372**	0.0%	372.0	6
n37-k6-60	**379**	0.0%	9	**379**	0.0%	379.0	5
n37-k6-45	**414**	0.0%	16	**414**	0.0%	414.0	6
n38-k5-90	**284**	0.0%	6	**284**	0.0%	284.2	5
n38-k5-60	**304**	0.0%	5	**304**	0.0%	304.0	5
n38-k5-45	**337**	0.0%	6	**337**	0.0%	337.0	5
n39-k5-90	**322**	0.0%	11	**322**	0.0%	322.0	5
n39-k5-60	**364**	0.0%	30	**364**	0.0%	364.0	7
n39-k5-45	**389**	0.0%	17	**389**	0.0%	389.0	6
n39-k6-90	303	2.0%	19	**303**	0.0%	303.2	4
n39-k6-60	**340**	0.0%	6	**340**	0.0%	340.0	5
n39-k6-45	**368**	0.0%	25	**368**	0.0%	368.0	6
n45-k6-90	368	0.5%	41	**368**	0.0%	368.8	7
n45-k6-60	**402**	0.0%	44	**402**	0.0%	402.0	7
n45-k6-45	**449**	0.0%	81	**449**	0.0%	449.0	7
n45-k7-90	456	0.2%	130	**456**	0.0%	456.4	11
n45-k7-60	479	0.2%	83	**479**	0.0%	480.6	10
n45-k7-45	500	0.2%	19	501	0.2%	501.0	10
n46-k7-90	357	0.3%	18	**357**	0.0%	360.2	9
n46-k7-60	**404**	0.0%	36	**404**	0.0%	404.0	9
n46-k7-45	**421**	0.0%	58	**421**	0.0%	421.0	8
n48-k7-90	415	0.2%	66	**415**	0.0%	415.4	10
n48-k7-60	**438**	0.0%	89	**438**	0.0%	438.6	8
n48-k7-45	**453**	0.0%	86	457	0.9%	458.0	8

Table 3.13 Exact and MSPT results: one category (set B)

Inst	CPLEX			MSPT			
Set B	BKS	%GAP LB	Time	Min	%BKS	Average	Time
n31-k5-90	**306**	0.0%	1	**306**	0.0%	306.0	8
n31-k5-60	**306**	0.0%	1	**306**	0.0%	306.0	6
n31-k5-45	**316**	0.0%	2	**316**	0.0%	316.0	5
n34-k5-90	**331**	0.0%	2	**331**	0.0%	331.0	7
n34-k5-60	**386**	0.0%	3	**386**	0.0%	386.0	6
n34-k5-45	**393**	0.0%	2	**393**	0.0%	393.0	7
n35-k5-90	**418**	0.0%	1	**418**	0.0%	418.0	7
n35-k5-60	**423**	0.0%	5	**423**	0.0%	423.0	4
n35-k5-45	**449**	0.0%	3	**449**	0.0%	449.0	4
n38-k6-90	330	0.3%	4	**330**	0.0%	330.0	7
n38-k6-60	**349**	0.0%	16	**349**	0.0%	349.0	8
n38-k6-45	**372**	0.0%	11	**372**	0.0%	372.0	6
n39-k5-90	**243**	0.0%	1	**243**	0.0%	243.0	5
n39-k5-60	**250**	0.0%	6	**250**	0.0%	250.0	4
n39-k5-45	**250**	0.0%	2	**250**	0.0%	250.0	4
n41-k6-90	**354**	0.0%	7	**354**	0.0%	354.0	7
n41-k6-60	**376**	0.0%	13	**376**	0.0%	376.0	8
n41-k6-45	436	0.2%	34	**436**	0.0%	436.0	8
n43-k6-90	**307**	0.0%	20	**307**	0.0%	307.0	7
n43-k6-60	**310**	0.0%	12	**310**	0.0%	310.0	7
n43-k6-45	341	0.3%	21	**341**	0.0%	341.0	8
n44-k7-90	405	0.2%	12	**405**	0.0%	405.0	12
n44-k7-60	429	0.2%	33	**429**	0.0%	429.0	11
n44-k7-45	430	0.2%	47	**430**	0.0%	430.0	9
n45-k5-90	**292**	0.0%	8	**292**	0.0%	292.0	7
n45-k5-60	**306**	0.0%	5	**306**	0.0%	306.0	6
n45-k5-45	**369**	0.0%	20	**369**	0.0%	369.0	7
n45-k6-90	**301**	0.0%	17	**301**	0.0%	301.0	9
n45-k6-60	315	0.3%	25	**315**	0.0%	315.0	9
n45-k6-45	**332**	0.0%	23	**332**	0.0%	332.0	8
n50-k7-90	**307**	0.0%	42	**307**	0.0%	307.2	9
n50-k7-60	**339**	0.0%	57	**339**	0.0%	339.0	8
n50-k7-45	**354**	0.0%	60	**354**	0.0%	354.0	9
n50-k8-90	558	0.2%	8	**558**	0.0%	558.0	20
n50-k8-60	558	0.2%	22	**558**	0.0%	561.4	18
n50-k8-45	606	0.2%	138	612	1.0%	613.8	14
n66-k9-90	556	10.4%	18.000	556	0.0%	556.8	21
n66-k9-60	571	14.7%	18.000	572	0.2%	572.6	18
n66-k9-45	577	8.3%	18.000	**577**	0.0%	577.8	16
n78-k10-90	511	64.8%	18.000	**508**	– 0.6%	508.0	25
n78-k10-60	519	49.7%	18.000	521	0.4%	523.6	21
n78-k10-45	554	65.5%	18.000	**553**	– 0.2%	555.4	22

Table 3.14 Summary of exact results: one category

Set	Opt	FS	n/a	Total
A	34	8	0	42
B	26	16	0	42
Total	60	24	0	84

In addition, it is important to show that the processing times with the exact method are competitive with the heuristic method. Even so, the processing time increases according to dimension of the instance, and, in the same two instances indicated above, where "% GAP LB" was high, the processing time has reached the time limit of 5 hours.

The result of the "% BKS" column makes it possible to show that MSPT generates high-quality solutions to the problem by considering only one category. In 89% of the instances a result is better or equal to the exact method and in the other cases (11% of instances) the percentage difference between the MSPT result and the exact method is no more than 1%.

Regarding the processing times, the highest value was 25 seconds (set n78-k10-90 instance A) and on average about 8 seconds.

Therefore, as detailed above, it is possible to obtain a solution with quality assurance in terms of the objective function and processing time using the exact method for most of the tested instances. Even so, the heuristic method was also able to generate high-quality solutions, even in the most difficult instances that were tested. Comparing the results of the exact method with those of the heuristic method in the one category problem helps to support the quality of the proposed heuristic for the six categories' problem which in many instances had a large "%GAP LB."

3.7.3 Comparison of the Total Distance Traveled with the Maximum Direct Distances on Routes

In this subsection we compare the objective of minimizing the classic routing problem (sum of the total distance traveled in each route) with the minimization objective considered in this search (sum of the maximum direct distance in each route) and the sum of the maximum direct distance in each route in the optimal solutions of the instances proposed by Augerat [1]. Thus, it was possible to compare the results of the optimal solutions of the instances proposed by Augerat [1] with the MSPT results in the same instances, only considering the 90-degree, 60-degree, and 45-degree threshold angles. The results for sets A and B are given in Tables 3.15 and 3.16, respectively. In each of these tables, the columns entitled "Opt Augerat" contain the results of the optimal solutions of the instances proposed by Augerat, and the column "Opt Augerat feasible?" indicates whether the optimal solution of the instances proposed by Augerat [1] is feasible considering the restrictions of the 90-degree, 60-degree, and 45-degree threshold angles. Below "Opt Augerat,"

Table 3.15 Results: distance traveled x maximum direct distance on each route (set *A*)

Inst	Opt Augerat			Opt feasible?			SMDD%			STD%			Add. vehicles		
Set A	SMDD	STD	Veh	90	60	45	90	60	45	90	60	45	90	60	45
n32-k5	325	784	5	S	N	N	−4.3%	0.0%	7.7%	12.5%	5.0%	6.5%	0	0	0
n33-k5	270	661	5	S	N	N	−4.4%	8.1%	23.0%	14.8%	15.1%	18.3%	0	1	2
n33-k6	304	742	6	N	N	N	−9.5%	7.6%	17.4%	14.2%	12.8%	10.6%	0	1	1
n34-k5	314	778	5	N	N	N	0.0%	5.4%	12.4%	19.0%	5.9%	6.9%	1	1	2
n36-k5	326	799	5	N	N	N	−4.9%	2.1%	6.4%	8.8%	7.6%	4.3%	0	1	1
n37-k5	249	669	5	N	N	N	−3.2%	16.1%	32.1%	10.3%	12.1%	16.3%	0	1	2
n37-k6	397	949	6	N	N	N	−6.3%	−4.5%	4.3%	13.2%	5.2%	6.1%	1	1	2
n38-k5	275	730	5	N	N	N	3.3%	10.5%	22.5%	13.6%	4.4%	10.7%	1	1	2
n39-k5	356	822	5	N	N	N	−9.6%	2.2%	9.3%	7.1%	11.1%	9.6%	1	2	3
n39-k6	321	831	6	N	N	N	−5.6%	5.9%	14.6%	8.1%	10.7%	8.5%	0	1	2
n45-k6	362	944	6	N	N	N	1.7%	11.0%	24.0%	15.6%	7.1%	11.4%	1	1	4
n45-k7	509	1146	7	S	N	N	−10.4%	−5.9%	−1.6%	17.6%	9.6%	5.9%	0	0	1
n46-k7	382	914	7	N	N	N	−6.5%	5.8%	10.2%	16.5%	13.3%	6.3%	0	1	1
n48-k7	482	1073	7	S	S	S	−13.9%	−9.1%	−5.2%	14.8%	13.9%	7.3%	0	0	0
Average							−5.3%	4.0%	12.7%	13.3%	9.6%	9.2%	0.4	0.9	1.6

Table 3.16 Results: distance traveled x maximum direct distance on each route (set B)

Inst	Opt Augerat			Opt feasible?			SMDD%			DT%			Add. vehicles		
Set B	SMDD	TD	Veh	90	60	45	90	60	45	90	60	45	90	60	45
n31-k5	316	672	5	S	S	S	−3.2%	−3.2%	0.0%	6.1%	4.3%	0.1%	0	0	0
n34-k5	346	788	5	S	N	N	−4.3%	11.6%	13.6%	10.4%	12.4%	10.0%	0	1	1
n35-k5	412	955	5	N	N	N	1.5%	2.7%	9.0%	10.9%	5.1%	5.3%	0	0	1
n38-k6	355	805	6	S	S	N	−7.0%	−1.7%	4.8%	11.6%	2.7%	3.0%	0	0	0
n39-k5	243	549	5	S	N	N	0.0%	2.9%	2.9%	1.3%	2.6%	1.6%	0	0	0
n41-k6	365	829	6	S	N	N	−3.0%	3.0%	19.5%	16.6%	4.1%	14.2%	0	1	2
n43-k6	328	742	6	N	N	N	−6.4%	−5.5%	4.0%	5.7%	4.7%	4.0%	0	0	1
n44-k7	414	909	7	N	N	N	−2.2%	3.6%	3.9%	10.2%	7.9%	8.0%	0	1	1
n45-k5	295	751	5	N	N	N	−1.0%	3.7%	25.1%	6.3%	2.3%	14.2%	1	1	3
n45-k6	297	678	6	N	N	N	1.3%	6.1%	11.8%	13.7%	13.0%	9.3%	1	1	1
n50-k7	330	741	7	S	N	N	−7.0%	2.7%	7.3%	14.7%	3.5%	3.4%	0	0	1
n50-k8	616	1312	8	S	N	N	−9.4%	−9.4%	−0.6%	12.3%	9.6%	6.6%	0	0	0
n66-k9	588	1316	9	S	S	S	−5.4%	−2.7%	−1.9%	10.6%	4.5%	4.6%	0	0	0
n78-k10	528	1221	10	N	N	N	−3.8%	−1.3%	4.7%	12.3%	4.7%	7.0%	0	0	0
Average							−3.6%	0.9%	7.4%	10.2%	5.8%	6.5%	0.1	0.4	0.8

the "SMDD" column indicates the sum of the maximum direct distance of each route, the "STD" column indicates the total distance of the routes, and the "Veh" column indicates the number of vehicles in the solution. Then, the heuristic results are presented using the MSPT in the same instances considering the threshold angles of 90 degrees, 60 degrees, and 45 degrees. The columns "SSMD%" indicate the percentage difference between the "Opt Augerat" result of the sum of the maximum direct distance of each route and the MSPT result. The "STD%" columns indicate the percentage difference between the "Opt Augerat" result of the sum of the total distances of the routes and the MSPT result. Finally, the "Additional vehicles" columns indicate the increase of vehicles between the "Opt Augerat" result of number of vehicles and the result of the MSPT.

When analyzing the information of the columns "Opt Augerat feasible" it is possible to notice that the smaller the threshold angle, the greater the number of non-viable Opt Augerat solutions. There are 57% of infeasible solutions when considering the 90-degree angle, 86% of infeasible solutions when considering 60 degrees, and 89% of infeasible solutions when considering 45 degrees. In addition, set B is the one that presents the largest number of non-viable solutions.

Regarding the "SMDD%" result all values are lower than those of the "Opt Augerat" solutions that are feasible. That is, the proposed heuristic method that considers the objective of minimizing the maximum direct distance of service of each route could generate solutions with a lower freight cost. The average percentage difference for the viable instances is − 6.6% considering the 90-degree threshold angle, −5.8% considering the threshold angle of 60 degrees, and −3.1% considering the 45-degree threshold angle. It is also possible to verify that the average percentage difference is greater for the instances of set A.

The results of the "SDT%" columns allow us to verify that the change in the minimization objective increases the sum of the distances of the routes that, on average, increases by 11.6%, that is, these solutions, although they result in a lower freight cost, have a higher variable cost of transportation. It is important to emphasize that because the "Opt Augerat" result does not consider the threshold angle constrain, routes with distant delivery points are formed, and therefore, this can bring practical problems to implement this solution with respect to the acceptance of this configuration by a carrier's driver, as discussed in Chap. 3.

With respect to the "Additional vehicles" results, the increase in the number of vehicles increases proportionally with the increase in the threshold angle. When considering the 90-degree threshold angle, only 20% of the instances had an increase of 1 vehicle, and when considering the threshold angle of 45 degrees, 75% of instances had an increase of 1 vehicle or more.

That is, based on the analysis of the results "% SDT" and "Increase of vehicles" it can be affirmed that when defining a very low threshold angle the fixed cost (number of vehicles) increases, but not necessarily this occurs with the variable cost (total distance traveled).

Therefore, as discussed in the above analysis, it is possible to observe a trade-off between freight cost and variable transport costs in relation to the objective of minimizing the problem.

3.8 Concluding Remarks

In this chapter we have addressed the problem of grouping orders into shipments with compatible categories and freight based on the direct distances to the farthest customers (PGOCCDD) that may arise in last mile distribution to nanostores in developing countries. We introduce two new attributes: compatibility of products that belong to different categories and the route cost determined by the maximum direct distance. We present a mathematical formulation for this problem and propose a heuristic based on GRASP, TS, and ILS to solve it.

Our results on extensive computational experiments show that our heuristic can effectively solve the PGOCCDD, as very good results could be obtained for the original and modified instances of sets *A* and *B* proposed by Augerat [1] for the VRP in short CPU times using an ordinary computer as opposed to the exact formulation, which struggled to reach good solution after long processing times (up to 5 hours) in a high-performance computer.

Based on the results of the proposed approach, it is possible to observe that assuming the freight to be paid on a route as the maximum direct distance a trade-off arises, comprising the freight cost and the variable transportation cost in relation to the minimization objective used. Solving the problem considering the minimization of the distance of the routes might generate higher freight costs, while solving the problem considering the minimization of the maximum direct distance generates an increase of the distance of the routes and, consequently, increase of the variable cost of transport.

The largest instance we solved (n78-k10 of set B) comprises 78 clients and totals 468 delivery orders when considering 6 product categories. One possible extension is to validate the proposed method with carrier data that provides service to wholesalers and/or retailers to evaluate the actual impact on the freight cost and the variable cost of the transport operation by changing the minimization objective of the classic VRP.

Another possible extension of the problem would be to consider an objective function with two hierarchical criteria: maximum direct distance and total distance traveled by the vehicles of the fleet. That is, it would be interesting to try to find solutions that, without changing the criterion given by the maximum direct distance, minimize as a secondary objective the total distance traveled.

Acknowledgments The authors acknowledge the use of computational facilities of LCCA-Laboratory of Advanced Scientific Computation of the University of São Paulo (LCCA-USP). The second author also acknowledges Brazil's CNPq (National Council for Scientific and Technological Development) for the financial support [grant number 306485/2023-0].

References

1. P. Augerat, *Approche polyèdrale du problème de tournées de véhicules*. Theses, Institut National Polytechnique de Grenoble—INPG (1995)
2. B.S. Everitt, S. Landau, M. Leese, D. Stahl, Cluster analysis, in *Wiley Series in Probability and Statistics*, 5 edn. (Wiley-Blackwell, Hoboken, NJ, 2011)
3. M.L. Fisher, R. Jaikumar, A generalized assignment heuristic for vehicle routing. Networks **11**, 109–124 (1981)
4. K. Fleszar, K.S. Hindi, New heuristics for one-dimensional bin-packing. Comput. Oper. Res. **29**(7), 821–839 (2002)
5. J.C. Fransoo, E.E. Blanco, C. Mejia-Argueta, *Reaching 50 Million Nanostores: Retail Distribution in Emerging Megacities* (2017)
6. M.R. Garey, D.S. Johnson, *Computers and Intractability—a Guide to the Theory of NP-Completeness* (W. H. Freeman and Company, New York, 1979)
7. M. Gendreau, G. Laporte, F. Semet, Heuristics and lower bounds for the bin packing problem with conflicts. Comput. Oper. Res. **31**(3), 347–358 (2004)
8. B.L. Golden, S. Raghavan, E.A. Wasil, The vehicle routing problem, in *Operations Research/-Computer Science Interfaces Series*, 2008 edn. (Springer, New York, 2008)
9. K. Jansen, J. Comb. Optim. **3**(4), 363–377 (1999)
10. K. Jansen, S. Öhring, Approximation algorithms for time constrained scheduling. Inf. Comput. **132**(2), 85–108 (1997)
11. L. Kaufman, P.J. Rousseeuw, *Finding Groups in Data: An Introduction to Cluster Analysis*, 9th edn. (Wiley, New York, 2009)
12. A. Khanafer, F. Clautiaux, E.-G. Talbi, New lower bounds for bin packing problems with conflicts. Eur. J. Oper. Res. **206**(2), 281–288 (2010)
13. Y.A. Koskosidis, W.B. Powell, Clustering algorithms for consolidation of customer orders into vehicle shipments. Trans. Res. Part B: Methodol. **26**(5), 365–379 (1992)
14. G. Laporte, M. Gendreau, J.-Y. Potvin, F. Semet, Classical and modern heuristics for the vehicle routing problem. Int. Trans. Oper. Res. **7**(4-5), 285–300 (2000)
15. J.M. Mulvey, M.P. Beck, Solving capacitated clustering problems. Eur. J. Oper. Res. **18**(3), 339–348 (1984)
16. A.E. F. Muritiba, M. Iori, E. Malaguti, P. Toth, Algorithms for the bin packing problem with conflicts. INFORMS J. Comput. **22**(3), 401–415 (2010)
17. I. Osman, N. Christofides, Capacitated clustering problems by hybrid simulated annealing and tabu search. Int. Trans. Oper. Res. **1**(3), 317–336 (1994)
18. H. Pirkul, Efficient algorithms for the capacitated concentrator location problem. Comput. Oper. Res. **14**(3), 197–208 (1987)
19. M. Quiroz-Castellanos, L. Cruz-Reyes, J. Torres-Jimenez, C.G.S. Héctor, J.F. Huacuja, A.C.F. Alvim. A grouping genetic algorithm with controlled gene transmission for the bin packing problem. Comput. Oper. Res. **55**, 52–64 (2015)
20. R. Sadykov, F. Vanderbeck, Bin packing with conflicts: A generic branch-and-price algorithm. INFORMS J. Comput. **25**(2), 244–255 (2013)
21. L.F.O.M. Santos, R.S. Iwayama, L.B. Cavalcanti, L.M. Turi, F.E. de Souza Morais, G. Mormilho, C.B. Cunha, A variable neighborhood search algorithm for the bin packing problem with compatible categories. Expert Syst. Appl. **124**, 209–225 (2019)
22. P. Toth, D. Vigo, *The Vehicle Routing Problem* (SIAM, New York, 2002)
23. T. Vidal, T.G. Crainic, M. Gendreau, C. Prins, Heuristics for multi-attribute vehicle routing problems: a survey and synthesis. Eur. J. Oper. Res. **231**, 1–21 (2013)
24. J. Baxter, Depot location: A technique for the avoidance of local optima. Eur. J. Oper. Res. **18**, 208–214 (1984). https://doi.org/10.1016/0377-2217(84)90186-3
25. T.A. Feo, M.G.C. Rezende, A probabilistic heuristic for a computationally difficult set covering problem. Oper. Res. Lett. **8**, 67–71 (1989)

26. B. Gillett, L. Miller, A heuristic algorithm for the vehicle dispatch problem. Oper. Res. **22**, 340–349 (1974). https://doi.org/10.1287/opre.22.2.340
27. F. Glover, Tabu Search - Part I. ORSA J. Comput. **1**, 190–206 (2006). https://doi.org/10.1287/ijoc.1.3.190
28. K. Hamdi-Dhaoui, N. Labadie, A. Yalaoui, The vehicle routing problem with conflicts. IFAC Proc. **44**, 9799–9804 (2011). https://doi.org/10.3182/20110828-6-IT-1002.01565
29. D.S. Johnson, Fast algorithms for bin packing. J. Comput. Syst. Sci. **8**, 272–314 (1974). https://doi.org/10.1016/S0022-0000(74)80026-7
30. S. Lin,. Computer solutions of the traveling salesman problem. Bell Syst. Tech. J. **44**, 2245–2269 (1965). https://doi.org/10.1002/j.1538-7305.1965.tb04146.x
31. I.H. Osman, Metastrategy simulated annealing and Tabu search algorithms for the vehicle routing problem. Ann. Oper. Res. **41**, 421–451 (1993). https://doi.org/10.1007/BF02023004
32. P. Toth, D. Vigo, The Granular Tabu search and its application to the vehicle-routing problem. INFORMS J. Comput. **15**, 333–346 (2003). https://doi.org/10.1287/ijoc.15.4.333.24890

Chapter 4
Mathematical Models and Algorithms for Large-Scale Transportation Problems

Carlos A. S. Oliveira

Abstract Intermodal transportation, also known as intermodal shipping or freight transport, involves moving goods using multiple transportation modes like trucks, trains, ships, and airplanes, with standardized containers facilitating seamless transfers without unloading and reloading cargo. This method offers several industrial benefits, including increased efficiency, cost savings, environmental advantages, and improved reliability. Efficiency is enhanced through reduced delays and minimized handling, while cost savings arise from optimizing various modes and routes. Environmentally, it is more sustainable, as trains and ships are more fuel-efficient and emit less pollution than long-haul trucking. Additionally, intermodal transportation reduces road congestion and highway wear by shifting long-haul transport to railways or waterways, and offers greater reliability by mitigating traffic and road closure impacts. In this paper, we consider a class of problems that occur by the interaction of two or more transportation modes, as used in intermodal transportation.

Keywords Optimization · Intermodal transportation · Routing efficiency · Machine learning

4.1 Introduction

Intermodal transportation, also known as intermodal shipping or intermodal freight transport, is a logistics and transportation concept that involves the movement of goods using multiple modes of transportation, such as trucks, trains, ships, and airplanes. The key feature of intermodal transportation is the use of standardized

---

C. A. S. Oliveira (✉)
Principal Inventive Scientist, Network Chief Technology Organization, AT&T Labs Inc., Bellevue, WA, USA
e-mail: carlos.oliveira@att.com

C. A. S. Oliveira, M. P. Pardalos (eds.), *Handbook of Artificial Intelligence and Data Sciences for Routing Problems*, Springer Optimization and Its Applications 219, https://doi.org/10.1007/978-3-031-78262-6_4

containers, which can be seamlessly transferred between different modes without the need to unload and reload the cargo.

There are several benefits in the industrial use of intermodal transportation, including efficiency, cost savings [12], environment benefits [10], and reliability, among others. First, considering efficiency, intermodal transportation is designed to be more efficient than relying solely on a single mode of transportation. It allows for the seamless transfer of cargo between different modes, reducing delays and minimizing handling.

Another important consideration is cost savings. By optimizing the use of various modes and routes, intermodal transportation can often be more cost-effective than using a single mode of transport for the entire journey. Environmental benefits are also important when considering intermodal transportation. Using this technique of transportation can be more environmentally friendly than long-haul trucking, as it often involves the use of trains or ships for the longest segments of the journey. These modes of transportation are generally more fuel-efficient and produce fewer emissions per unit of cargo.

In terms of network congestion, we can see a reduction of costs by shifting a portion of long-haul transportation from roads to railways or waterways. Thus, intermodal transportation can help reduce road congestion [2] and wear and tear on highways.

Finally, one can consider the reliability of this type of transportation. Intermodal techniques can offer greater reliability and predictability, as it is less affected by traffic congestion or road closures than long-haul trucking.

An important aspect of intermodal transportation is its relationship to global trade [11]. Intermodal is a crucial factor for international trade, as standardized containers make it easier to move goods between countries and continents. Preparing for intermodal makes it possible to interact with transportation systems available in different countries and geographies.

Typically, a shipment in intermodal transportation starts with the cargo being loaded into standardized containers, which are then transported by trucks to an intermodal terminal. At the terminal, the containers are transferred to trains or ships for long-distance transport. Finally, they may be transferred back to trucks for delivery to their final destination. The use of standardized containers and well-established procedures makes these transfers efficient and reduces the risk of damage or loss during transportation.

Intermodal transportation is widely used for a variety of goods, including consumer products, raw materials, and industrial equipment. It plays a crucial role in the global supply chain and logistics industry [3].

Typical examples of transportation modes include:

- From land (trucks) to airplanes
- Naval ships to land transportation (trucks)
- From land (trucks) to land (railways)

In this paper, we consider a class of problems that occur by the interaction of two or more transportation modes, as used in intermodal transportation.

4.1.1 Challenges of Multimodal Transportation

Multimodal transportation, which involves the use of multiple modes of transport (such as road, rail, sea, and air) for moving goods and people, offers numerous advantages, but it also comes with its share of challenges. Some of the key challenges associated with multimodal transportation are discussed in this section.

One of the primary challenges is coordinating and integrating the various modes of transportation involved. Different modes may have separate operators, schedules, regulations, and infrastructure, which can lead to inefficiencies and disruptions in the supply chain [12]. Ensuring that different modes of transportation are compatible in terms of infrastructure (e.g., container sizes, loading/unloading facilities) can be challenging. Incompatibilities can lead to delays, additional handling, and increased costs.

Regulatory and legal hurdles are also part of the challenge. Different modes of transportation are subject to various regulations and legal requirements. Navigating these regulations, including customs, safety, and environmental standards, can be complex and time-consuming [2]. Transferring cargo from one mode of transportation to another can introduce risks of damage, loss, or theft. Handling and transfer procedures need to be carefully managed to minimize these risks.

Security is a significant concern in multimodal transportation, particularly for international shipments. Cargo may be vulnerable to theft or tampering during transfers between modes, and security measures need to be in place to protect shipments [10]. Delays can occur at various points in the multimodal transportation process, including congestion on roads, port or terminal congestion, and transportation breakdowns. These delays can disrupt supply chains and increase costs.

Here are some additional issues that impact multimodal transportation:

- Environmental Impact: While multimodal transportation can be more environmentally friendly than relying solely on road transport, each mode of transportation still has its environmental impact. Reducing this impact and ensuring environmentally sustainable practices can be a challenge [11].
- Cost Management: Coordinating multiple modes of transportation can sometimes lead to higher administrative and operational costs. Ensuring that the cost structure is competitive and that cost efficiencies are realized can be challenging [3].
- Liability and Insurance: Determining liability in the event of cargo damage or loss can be complex when multiple parties and modes are involved. It's important to have clear liability agreements and appropriate insurance coverage.
- Supply Chain Visibility: Maintaining visibility and tracking of cargo throughout its journey in a multimodal supply chain can be challenging. It's important for companies to have real-time information about the status and location of their shipments.
- Technological Integration: Implementing technology systems that can seamlessly manage and integrate information across multiple modes of transportation is a

technological challenge. Data sharing and system compatibility are crucial for smooth operations.
- Weather and Natural Disasters: Multimodal transportation can be vulnerable to weather-related disruptions and natural disasters, especially for modes like air and sea transport. Planning for contingencies and risk management is essential.

Despite these challenges, multimodal transportation remains a valuable strategy for optimizing supply chain efficiency and reducing transportation costs, particularly in long-distance and international logistics. Effective management, coordination, and risk mitigation strategies are essential to overcome these challenges and realize the benefits of multimodal transportation.

4.1.2 Truck/Rail Intermodal

Truck/rail intermodal transportation, often referred to as intermodal shipping, is a method of transporting freight using a combination of trucks and trains. It leverages the strengths of both modes of transportation to offer a cost-effective and efficient way to move cargo over long distances.

The process begins with the cargo being loaded into standardized containers, typically 20 feet or 40 feet in length. These containers are designed to be easily transferable between trucks and railcars without the need to unpack and repack the cargo. The use of standardized containers is a key feature of intermodal transportation.

The cargo containers are initially transported by trucks to an intermodal terminal or ramp. This terminal is typically strategically located near the origin or destination of the cargo, but it can also be an intermediate point.

At the intermodal terminal, the cargo containers are lifted onto specially designed railcars. This transfer can be done using specialized equipment such as cranes or straddle carriers. The containers are securely mounted on the railcars for the train journey. Next, the loaded railcars, carrying the cargo containers, are moved by train to the intermodal terminal nearest to the cargo's destination. Trains are well-suited for long-distance transportation because they are fuel-efficient and capable of carrying a large volume of cargo.

At the destination intermodal terminal, the cargo containers are unloaded from the railcars and transferred back onto trucks. Again, specialized equipment is used to lift the containers from the railcars and place them on the trucks.

The next step is final delivery. The cargo containers are then transported by trucks or smaller vehicles to their final destination. The destination could be a distribution center, warehouse, or directly to the customer. This is typically a short-haul or "last-mile" trucking leg.

4.1.3 Benefits of Truck/Rail Intermodal

There are several advantages in using this intermodal transportation system, including the following:

- Efficiency: Intermodal transportation is designed to be more efficient than long-haul trucking alone. Trains are particularly efficient for transporting large volumes of cargo over long distances, while trucks are well-suited for the flexibility and accessibility of short-distance transportation.
- Cost Savings: By leveraging the cost advantages of rail transportation for the long-haul portion of the journey, intermodal transportation can often be more cost-effective than relying solely on trucks. Reduced Road Congestion: Shifting cargo from trucks to trains for the long-distance portion of the journey helps reduce road congestion and wear and tear on highways.
- Environmental Benefits: Trains are generally more fuel-efficient and produce fewer emissions per unit of cargo than trucks, making intermodal transportation more environmentally friendly for long-distance shipping. Truck/rail intermodal transportation is widely used in the freight industry for various types of cargo, including consumer goods, raw materials, and industrial products. It plays a significant role in the global supply chain, helping to move goods efficiently and sustainably over long distances.

4.1.4 Main Intermodal Concepts

The process of truck/rail intermodal transportation involves several key actors or parts, each with distinct roles and responsibilities to ensure the smooth movement of cargo from origin to destination. Here are the main actors and parts involved in the truck/rail intermodal process:

- Shipper: The shipper is the entity or organization that owns or produces the goods being transported. They are responsible for arranging the transportation of their cargo and preparing it for shipment, including packing and loading it into containers.
- Intermodal Terminal: Intermodal terminals serve as crucial hubs in the intermodal transportation process. These facilities are equipped with infrastructure for transferring cargo between trucks and railcars. They facilitate the transfer, storage, and organization of containers. Intermodal terminals can be located at the point of origin or destination or at intermediate points along the transportation route.
- Truck Carrier: Truck carriers, also known as drayage carriers, provide the road transport component of the intermodal process. They pick up cargo containers from the shipper's location or the origin intermodal terminal and deliver them to

the rail terminal for transfer to railcars. At the destination, they receive containers from the rail terminal and transport them to their final destination.

- Rail Carrier: Rail carriers operate the trains that transport the cargo containers over long distances. They are responsible for moving the loaded containers from the origin rail terminal to the destination rail terminal. Rail carriers are typically large freight railroad companies.
- Containers: Standardized containers, typically 20 feet or 40 feet in length, are used to package and transport the cargo. These containers are designed to be easily transferred between trucks and railcars. They come in various types, including dry van containers, refrigerated containers, and specialized containers for specific cargo types.
- Chassis: Chassis are specialized trailers designed to transport cargo containers. Truck carriers use chassis to transport containers to and from the rail terminals. Chassis come in different types, depending on the container size and design.
- Cranes and Handling Equipment: Intermodal terminals are equipped with cranes, straddle carriers, or other handling equipment to lift and transfer containers between trucks and railcars. These machines are essential for efficient cargo handling at terminals.
- Customs and Regulatory Authorities: Various government agencies are involved in regulating and overseeing international intermodal shipments. Customs agencies ensure compliance with import and export regulations, collect duties and taxes, and inspect cargo for security and safety.
- Technology and Tracking Systems: Information technology and tracking systems are used to manage and track cargo throughout the intermodal journey. This includes tracking the location, condition, and status of cargo containers, as well as managing bookings, scheduling, and documentation.
- Third-Party Logistics Providers (3PLs): Some shippers and cargo owners may use third-party logistics providers to manage and coordinate the entire intermodal transportation process. 3PLs can help with route planning, booking, documentation, and other logistics services.

The successful coordination and collaboration of these actors and components are critical to the efficient and reliable operation of truck/rail intermodal transportation. This mode of transportation relies on a well-organized supply chain to move goods effectively across different modes of transport and provide cost-effective and environmentally friendly solutions for long-distance shipping.

4.2 Intermodal Truck/Rail Process

The process of intermodal transportation involves several key activities, beginning with the pickup of loads at their origin locations. This initial stage requires coordination and planning to ensure that the goods are collected efficiently from various sites. Specialized equipment, such as trucks or other vehicles, is typically used to gather the cargo from warehouses, manufacturing facilities, or other points of origin [13]. The focus here is on ensuring that the loads are securely packed and ready for transport to the next stage without any delays or damage.

Once the loads are picked up, they are transported to the origin rail yard. This leg of the journey involves moving the cargo from the initial pickup points to a centralized rail facility where they can be consolidated and prepared for the longer rail journey. This stage requires careful logistical planning to optimize routes and ensure timely delivery to the rail yard. Efficient coordination with local transportation providers is crucial to minimize transit times and costs [6].

The core of the intermodal journey is the rail leg, executed by the railroad. At this stage, the cargo is loaded onto trains using standardized containers that facilitate easy handling and transfer. The rail segment is typically the longest part of the journey and leverages the efficiency and capacity of rail transport to move large volumes of goods over long distances. Rail transport is generally more fuel-efficient and environmentally friendly than road transport, making it a preferred option for the bulk of the journey [18]. This stage involves close coordination with the railroad operators to ensure schedules are met and any potential delays are managed effectively.

Upon reaching the destination rail yard, the next step is to transport the loads from the rail yard to their final destination. This involves unloading the containers from the trains and transferring them to trucks or other vehicles for the final leg of the journey. The final delivery requires precise timing and coordination to ensure that the goods reach their end-users or distribution centers as planned [9]. This stage often involves navigating urban or regional road networks, requiring detailed route planning and timing to avoid congestion and ensure timely deliveries.

Finally, after the loads have been delivered, the process involves repositioning the transportation equipment, such as trucks or containers, to the next origin point. This stage ensures that the transportation assets are efficiently utilized and prepared for the next cycle of pickups and deliveries. Repositioning involves planning and logistics to optimize the return routes and minimize empty miles, thereby improving overall efficiency and reducing operational costs. This continuous cycle of repositioning and readiness is essential for maintaining the seamless flow of goods in the intermodal transportation network.

4.3 Intermodal Routing Problem

Determining an optimal way for an intermodal carrier (IMC) to serve a given set of loads involves complex logistical and operational decisions. The primary challenge is to efficiently coordinate the transportation of goods using multiple modes, including trucks, trains, and possibly ships, while considering the costs associated with each mode. The IMC must decide how to route the loads, which vendors to use, and when it might be more cost-effective to off-load certain loads to other vendors. These decisions are influenced by various factors, including the cost structure of each mode, the availability and capacity of transportation resources, and the specific requirements of the loads being transported.

In addition to routing and vendor selection, the IMC must also manage the timing and scheduling of each leg of the journey to ensure timely delivery and minimize delays. This involves synchronizing the different modes of transportation and managing the transfer points where goods are moved from one mode to another. The problem is further complicated by the need to maintain service quality and reliability while minimizing costs. Given the dynamic nature of transportation networks and the variability in demand and supply, the IMC must be adaptable and responsive to changing conditions.

4.3.1 Problem Objective

The objective of this optimization problem is to minimize the operating costs associated with transporting the given set of loads. Several key cost components need to be considered:

Fuel costs are a significant expense in intermodal transportation. These costs vary depending on the distance traveled, the mode of transportation, and current fuel prices. Efficient route planning and mode selection can help reduce fuel consumption and, consequently, fuel costs.

The railroad contract cost is another major expense. This includes the fees paid to railroad companies for the transportation of goods over long distances. Negotiating favorable contract terms and optimizing the use of rail transport can help manage these costs.

Drayage costs refer to the expenses associated with using trucks for the short-distance transportation of goods to and from rail yards and other transfer points. These costs can be minimized by optimizing the scheduling and routing of trucks to reduce empty miles and improve utilization.

Container costs include the expenses related to the use of standardized containers for transporting goods. These costs can vary depending on the duration of use, the type of container, and any associated fees. Efficient container management, including timely returns and minimizing idle time, can help reduce these costs.

Chassis costs are related to the use of chassis for transporting containers by road. Similar to container costs, these expenses can be managed by optimizing the utilization and minimizing the downtime of chassis.

By carefully considering and optimizing these cost components, the IMC can develop an effective transportation plan that minimizes operating costs while ensuring timely and reliable delivery of loads. This requires a comprehensive approach that integrates route planning, mode selection, vendor management, and cost control strategies.

4.4 Mathematical Model

To create a mathematical model for multimodal routing between rail and trucks that determines the minimum cost route for moving products from a set of origins to a set of destinations with required demands, you can formulate this as a mixed-integer linear programming (MILP) problem. In this model, we consider decision variables, objective function, and constraints.
Sets:

- O: Set of origins
- D: Set of destinations
- OD: Set of origin-destination pairs

Parameters:

- C[i,j]: Cost of transporting one unit of product from origin i to destination j using the multimodal system
- R[i,j]: Cost of transporting one unit of product from origin i to destination j using a truck
- T[i,j]: Cost of transporting one unit of product from origin i to destination j using a rail
- D[od]: Required demand for origin-destination pair od
- R_cap[i]: Capacity of trucks departing from origin i
- T_cap[i]: Capacity of trains departing from origin i

Decision Variables:

- X[od]: Binary variable denoting whether to use the multimodal system for the od origin-destination pair (1 if used, 0 if not)
- Y[od]: The number of units of product to transport from origin o to destination d using the multimodal system (continuous variable)

Objective Function: Minimize the total transportation cost:

$$\min Z = \sum_{od \in OD} C_{od} Y_{od} + \sum_{od \in OD} X_{od} (R_{od} D_{od} - T_{od} Y_{od}) \tag{4.1}$$

Subject to:
Demand Fulfillment: Ensure that the demand for each origin-destination pair is met.

$$\sum_{od \in OD} Y_{od} = D_{od} \forall od \in OD \tag{4.2}$$

Capacity Constraints: Ensure that the capacity of trucks and trains at each origin is not exceeded.

$$\sum_{d \in D} X_{od} Y_{od} \leq Rcap_o \forall o \in O, d \in D \tag{4.3}$$

Binary Variable Constraints: These constraints ensure that if X[od] is 1, then Y[od] must be greater than or equal to 0.

$$Y_{od} \geq 0 \forall o, d \in O, D \tag{4.4}$$

Non-negativity of Y: The number of units transported using the multimodal system must be non-negative.

$$X_{od} \in \{0, 1\} \forall o, d \in O, D \tag{4.5}$$

This mathematical model represents a MILP that aims to minimize the total transportation cost while satisfying demand requirements and capacity constraints for both trucks and trains. The binary variable X[od] is used to determine whether the multimodal system is used for each origin-destination pair, and Y[od] represents the quantity of products transported. Solving this model will provide the optimal routes and quantities for each origin-destination pair.

4.5 Using IM Loops

One of the largest costs associated with intermodal transportation (IM) is the ownership and use of equipment, such as trailers and containers. The expense of acquiring, maintaining, and operating this equipment can be substantial, making it imperative to utilize these assets as efficiently as possible. To address this, a general strategy that employs the use of loops is adopted.

The loop strategy focuses on improving the utilization of expensive equipment. By ensuring that all equipment returns to its origin after completing a series of transportation legs, the strategy maximizes the use of each asset and minimizes the need for additional equipment. This not only reduces costs but also streamlines operations, as equipment is consistently cycled through a predictable and efficient route. The continuous return to origin simplifies logistical planning and ensures that assets are available where they are needed most.

4.6 Solution Format

Each part of an intermodal loop is composed of four transportation legs. The first leg involves trucking the load from the origin to a rail yard, a process known as drayage. This step is crucial for moving goods from their initial location to the point where they can be loaded onto trains for the longer haul. The second leg involves rail transport from the origin yard to the destination yard. This segment leverages the efficiency and cost-effectiveness of rail transport for covering long distances.

Upon reaching the destination yard, the third leg involves drayage once again, this time transporting the load from the destination yard to its final destination. Finally, the fourth leg involves repositioning the truck from the final destination back to the next origin point. This step ensures that the truck is ready for the next load, maintaining the efficiency of the overall loop. A loop is thus a sequence of two or more such routes, connected by repositioning steps as necessary to ensure continuous operation.

4.7 Solution Costs

The costs associated with intermodal loops include several key components. Fuel costs are a significant expense, influenced by the distance traveled and the mode of transportation used. Rail contract costs, which are typically distance-based, represent another major component. These costs are associated with the fees paid to railroad companies for the use of their infrastructure and services.

Drayage costs involve the expenses related to the short-distance transportation of goods to and from rail yards. These costs can vary depending on the distance and time required for these transfers. Driver costs are also an important consideration, encompassing wages, benefits, and other expenses related to employing drivers for the trucking legs of the journey. Lastly, equipment utilization costs include the expenses related to owning, maintaining, and operating trailers and containers. Efficient management and utilization of this equipment are crucial for minimizing these costs and ensuring the overall cost-effectiveness of the intermodal transportation system.

4.8 Problem Decomposition

The structure of the intermodal transportation optimization problem lends itself well to an algorithmic decomposition strategy. This approach breaks down the complex, large-scale problem into more manageable subproblems. The first step in this strategy involves using an algorithm to generate loops that encompass each transportation lane. By creating these loops, the algorithm identifies potential routes that can be utilized for moving goods efficiently. This step ensures that all possible pathways are considered, providing a comprehensive set of options for the subsequent optimization process.

Once the loops are generated, the next step is to apply a mathematical programming formulation to solve the problem of finding the most cost-effective routes. This involves using optimization techniques to evaluate the generated loops and select the routes that minimize operating costs while meeting all logistical constraints. The mathematical model considers various factors, including fuel costs, rail contract costs, drayage, driver expenses, and equipment utilization. By integrating these elements into a cohesive optimization framework, the IMC can identify the cheapest and most efficient routes for transporting goods, thereby achieving significant cost savings and operational efficiency.

The algorithm for building single routes is designed to efficiently determine a set of paths that connect loads from one location to another, incorporating intermediate stops if necessary.

Here is the algorithm to build single routes:

- Set of paths $P \leftarrow \emptyset$
- **For each** load from locations a to b **do**:
- check rail locations O close to a
- check rail locations D close to b
- generate IM paths from a to O to D to b.
- add to P the α best such paths (α is a parameter).
- **End-For**
- create loops of size at most K (usually $K \leq 3$) using the members of P.
- stop when all paths have been covered, or when remaining paths are all disjoint.

The procedure above will create a set of loops R that cover the loads and can be used to minimize cost in the next step. The process begins with initializing an empty set of paths P. For each load requiring transportation from an origin a to a destination b, the algorithm first identifies nearby rail locations, O and D, which are proximate to a and b, respectively. These rail locations serve as potential transfer points in the route.

Once the relevant rail locations are identified, the algorithm generates intermodal (IM) paths that traverse from a to O, then from O to D, and finally from D to b. Among the generated paths, the algorithm selects the top α paths based on a predetermined criterion, where α is a configurable parameter reflecting the number of paths to consider for each load. These selected paths are then added to the set P.

Following the path generation and selection, the algorithm proceeds to create loops of a specified maximum size K (typically, $K \leq 3K \leq 3$). These loops are constructed using the paths in P, ensuring that they cover as many loads as possible. The creation of loops continues until all paths have been accounted for, or until the remaining paths are all disjoint, meaning no further connections can be made.

The resultant set of loops R produced by this algorithm is intended to comprehensively cover the loads, facilitating an efficient transport network. These loops can then be employed in subsequent steps to optimize and minimize transportation costs, leveraging the structured path and loop framework established in this algorithm.

4.9 Loop Selection Formulation

Another useful formulation for this problem is given as follows. Consider the following decision variables:

- $x_i \in Z$: number of loads served by loop $i \in R$
- y_l: the number of loads served by third-party IMCs for lane $l \in L$ (non-economical lanes)

Here is some additional notation used:

- R: set of loops.
- L: set of lanes.
- $u(l, i) = 1$ if lane l is traversed by loop i.
- N_C: number of containers available.
- N_T: number of trailers available.
- c_i: cost of loop i
- c'_l: cost of servicing lane l with 3P IMCs.
- q_l: number of loads serviced by lane l.

The loop selection formulation can be defined as follows:

$$\min \sum_{i \in R} c_i x_i + \sum_{l \in L} c'_l y_l$$

Subject to:

$$\sum_{l:u(l,i)=1} x_i + y_l \geq q_l \qquad \forall l \in L \quad \text{(coverage constr.)}$$

$$\sum_{i \in R} x_i \leq N_C + N_T \qquad \text{(capacity constraint)}$$

$$y_l, x_i \in Z, \qquad \forall l \in L, \forall i \in R$$

The loop selection formulation is a mathematical model designed to optimize the selection of transportation loops and additional paths to minimize costs while meeting specific constraints. This formulation involves minimizing a combined cost function and adhering to coverage and capacity constraints. Here's a breakdown of the formulation in plain English:

The objective of the loop selection formulation is to minimize the total cost, represented by:

$$\min \sum_{i \in R} c_i x_i + \sum_{l \in L} c'_l y_l$$

- $\sum_{i \in R} c_i x_i$: This term represents the total cost associated with selecting specific loops i from the set R. Here, c_i is the cost of selecting loop i, and x_i is a binary variable indicating whether loop i is selected (1) or not (0).
- $\sum_{l \in L} c'_l y_l$: This term represents the total cost associated with additional paths l from the set L. Here, c'_l is the cost of selecting additional path l, and y_l is a binary variable indicating whether path l is selected (1) or not (0).

1. **Coverage Constraint**:

$$\sum_{l:u(l,i)=1} x_i + y_l \geq q_l \qquad \forall l \in L$$

   This constraint ensures that each load l in the set L is adequately covered. The function $u(l, i)$ indicates whether loop i includes load l (1 if true, 0 if false). The sum of x_i for all loops i covering load l and the additional path y_l must be at least q_l, ensuring sufficient coverage for each load.
2. **Capacity Constraint**:

$$\sum_{i \in R} x_i \leq N_C + N_T$$

   This constraint limits the total number of selected loops to the combined capacity of N_C (the number of available cars) and N_T (the number of available trains). It ensures that the selected loops do not exceed the available transportation capacity.
3. **Binary Variables**:

$$y_l, x_i \in \{0, 1\}, \qquad \forall l \in L, \forall i \in R$$

   This condition states that the variables y_l and x_i must be binary (0 or 1). This means that each loop i and additional path l is either selected (1) or not (0).

The loop selection formulation aims to find the optimal combination of transportation loops and additional paths to minimize costs while ensuring all loads are covered and the transportation capacity is not exceeded. This is achieved by selecting the appropriate x_i and y_l values under the given constraints.

4.9.1 Loop Selection Evaluation

In this section, we consider the results of the loop section model. Notice that this is a large-scale mixed-integer programming (MIP) problem, involving millions of variables and constraints.

However, due to its structure, FICO Xpress can find solutions very close to the optimum (within a 0.1% optimality gap) in a reasonable time frame (less than 1 hour) when utilizing 256 GB of memory.

By leveraging the capabilities of FICO Xpress, we decomposed this very complex problem into manageable subproblems that can be solved on hardware readily available on Amazon Web Services (AWS).

The practical application of this model has led to significant cost savings, with an estimated $1.27 million saved in operating costs for a large USA-based transportation unit, during its first 2 years of implementation.

The evaluation of the loop selection model highlights its effectiveness and efficiency in solving large-scale optimization problems. Despite the immense complexity and size of the MIP, the structured approach allows FICO Xpress to achieve near-optimal solutions swiftly. This rapid solution process is crucial for real-time decision-making and operational efficiency.

By decomposing the overarching problem into smaller, more manageable subproblems, we were able to utilize AWS's readily available hardware resources effectively. This approach not only made the problem more tractable but also ensured that the solutions could be obtained within practical time frames and computational limits.

The financial impact of this model is substantial. Over the first 2 years of its deployment, the model has contributed to saving approximately $1.27 million in operating costs. These savings underscore the value of implementing advanced optimization techniques in operations research, demonstrating tangible benefits in terms of cost reduction and resource optimization.

4.10 Using a GA to Solve the Problem

Using a MIP to solve the multimodal problem is possible, but sometimes we need a faster way to evaluate a certain scenario. In this situation, it is useful to use a metaheuristic such as GA to solve the problem [17].

Implementing a genetic algorithm for the multimodal routing problem involves coding the GA operations (selection, crossover, mutation), defining the fitness function, and handling constraints related to capacity and route feasibility. The algorithm can be fine-tuned to balance exploration and exploitation for efficient convergence to optimal or near-optimal solutions.

A genetic algorithm (GA) is a metaheuristic optimization technique inspired by the process of natural selection and evolution. GAs are well-suited for solving complex optimization problems like the multimodal routing problem [16].

In this section, we provide a high-level description of how to use a genetic algorithm to solve this multimodal problem. First, we have the initialization phase. In this phase we create an initial population of potential solutions (chromosomes). Each chromosome represents a potential routing solution from origins to destinations.

Next, we encode the chromosomes to represent routes, for example, using a binary string or permutation encoding.

The algorithm also uses fitness evaluation. Fitness evaluation in GA involves calculating the total cost of routing solutions, considering transportation costs and capacity constraints [14]. For this, we calculate the fitness of each chromosome in the population. The fitness function should represent the total cost of the routing solution, which is the objective to minimize. The fitness function takes into account the cost of transportation for each route, capacity constraints, and other relevant factors.

An important step is the selection mechanism. GA uses selection mechanisms to propagate chromosomes with higher fitness, simulating natural selection [5]. We use a selection mechanism to choose chromosomes from the current population to form a new generation. Chromosomes with better fitness have a higher chance of being selected, simulating the principle of "survival of the fittest."

4.10.1 GA for Loop Selection Problem

Designing a genetic algorithm (GA) to solve the loop selection problem involves encoding potential solutions as chromosomes, defining fitness functions, and establishing genetic operators for selection, crossover, and mutation. Here's a step-by-step outline of how the genetic algorithm can be designed:

1. **Encoding the Chromosome**: Each chromosome represents a possible solution, i.e., a selection of loops and additional paths. We'll use binary encoding where:
   - x_i and y_l are binary variables (0 or 1).
   - The chromosome will be a binary string of length $|R| + |L|$, where the first $|R|$ bits represent x_i values and the next $|L|$ bits represent y_l values.
2. **Initial Population**: Generate an initial population of chromosomes randomly or using a heuristic approach to ensure diversity.
3. **Fitness Function**: The fitness of each chromosome is determined by how well it minimizes the total cost while satisfying the constraints. The fitness function can be defined as:

$$\text{Fitness}(chromosome) = \left(\sum_{i \in R} c_i x_i + \sum_{l \in L} c'_l y_l \right) + \text{Penalty}$$

   where penalty is a large value added if the chromosome violates any of the following constraints:

   - **Coverage Constraint**:

$$\sum_{l:u(l,i)=1} x_i + y_l \geq q_l \qquad \forall l \in L$$

- **Capacity Constraint**:

$$\sum_{i \in R} x_i \leq N_C + N_T$$

4. **Selection**: Use a selection method like roulette wheel selection or tournament selection to choose parent chromosomes based on their fitness.
5. **Crossover**: Perform crossover (e.g., single-point or two-point crossover) to combine pairs of parent chromosomes and produce offspring. This operation combines segments of two parent chromosomes to create new chromosomes.
6. **Mutation**: Apply mutation to some bits of the chromosomes to maintain genetic diversity and explore new solutions. For binary encoding, this can be done by flipping bits from 0 to 1 or 1 to 0.
7. **Constraint Handling**: Ensure that the offspring chromosomes respect the constraints. If a chromosome violates any constraints, apply repair mechanisms or impose penalties in the fitness function.
8. **Replacement**: Replace a portion of the population with the new offspring to form the next generation.
9. **Termination**: Continue the evolution process for a fixed number of generations or until convergence criteria (e.g., no improvement in fitness) are met.

4.11 Transportation Mode Selection

In the context of intermodal transportation, the selection of the appropriate mode for a given load is a critical decision that impacts both efficiency and cost. The problem involves the consideration of thousands of truck loads each week, requiring a strategic approach to match demand with available supply. The available supply is defined by the number of containers and drivers, each with specific characteristics that influence their suitability for different types of loads. Effective transportation mode selection ensures that each load is transported in the most efficient and cost-effective manner, considering the unique requirements and constraints of the logistics operation.

The demand for transportation services must be met with an optimal allocation of resources. Containers and drivers are the two primary resources that need to be managed. Containers come in various sizes, and each size has its own set of features that determine its suitability for different types of loads. Similarly, drivers can be categorized based on their driving configurations, such as solo drivers or team drivers. Each category has different operational characteristics, including the distances they can cover, the regulations they must comply with, and the associated costs. These characteristics must be carefully considered to ensure that the transportation mode selected for each load maximizes efficiency while minimizing costs.

4.12 Transportation Selection Options

The options available for transportation mode selection are diverse, reflecting the varied needs of the logistics operation. Container sizes, for example, range from 32 inches to 53 inches. Each size offers different capacities and is suitable for different types of loads. Smaller containers might be more appropriate for lighter, less bulky goods, while larger containers are ideal for heavier, bulkier items. The selection of the appropriate container size is crucial for optimizing the use of available space and minimizing transportation costs.

Driver combinations are another critical factor in transportation mode selection. Solo drivers operate individually, and their operations are subject to specific legislation and cost structures. They might be suitable for shorter distances or routes that do not require continuous driving. On the other hand, team drivers, who operate in pairs, can cover longer distances more quickly due to their ability to drive continuously by alternating shifts. This configuration is ideal for long-haul routes where speed and efficiency are paramount. However, team drivers also come with higher costs and different regulatory requirements. By considering these factors, logistics managers can make informed decisions that balance cost, efficiency, and compliance with regulations, ensuring that each load is transported in the best possible manner.

Determining the best transportation mode for each load in a given set of loads is a complex problem that requires careful consideration of various factors. Each load comes with a unique set of features that can significantly impact the decision-making process. These features include the origin and destination of the load, which determine the distance and potential routes available. The cost associated with transporting the load is another critical factor and is often a continuous variable that needs to be minimized. Additionally, the priority of the load—whether it is express, normal, or low-priority—affects how quickly it needs to be delivered and, consequently, the transportation mode chosen. The owner of the load might also have specific preferences or contractual agreements that influence the decision. Finally, the day of the week can play a role, as certain days might have different traffic patterns, availability of transportation modes, or cost implications.

Furthermore, there is a wealth of historical data that can provide insights into mode selection and associated costs for each load. This data encompasses previous decisions made by managers, the outcomes of those decisions, and the costs incurred. By analyzing this data, patterns and trends can be identified that help in predicting the most cost-effective and efficient transportation modes for future loads. The traditional approach relies heavily on the experience and intuition of managers, who make decisions based on available supply and their past experiences. While this method has its merits, it may not always yield the optimal results, especially given the complexity and variability of the factors involved.

To improve the decision-making process, advanced data analytics and optimization algorithms can be employed. These tools can analyze the vast amounts of historical data and use machine learning techniques to predict the best transportation mode for each load. By considering all relevant features and learning from past decisions, these algorithms can provide recommendations that are more accurate and cost-effective than those based solely on human intuition. This approach not only leverages the wealth of available data but also ensures that decisions are consistent and based on objective criteria. Implementing such a system can lead to significant improvements in efficiency and cost savings, ultimately enhancing the overall effectiveness of the transportation process.

4.13 ML for Transportation Mode Selection

The approach to determining the best mode of transportation can be effectively tackled using machine learning (ML) techniques. This problem can be viewed as a classification task where the transportation mode is the assigned label. By treating the selection of the transportation mode as a classification problem, we can leverage advanced ML algorithms to predict the most suitable mode based on various features associated with each load.

A random forest classifier was employed to learn the labels for several years of test data. Random forest is known for its excellent performance in classification tasks, making it comparable to other robust methods such as support vector machines. The classifier was trained using historical data, enabling it to identify patterns and make accurate predictions for new loads. The effectiveness of random forest in this context is supported by research, such as the study by Hagenauer and Helbich [7], which conducted a comparative analysis of seven ML classifiers for modeling travel mode choice. Their study showed that advanced classifiers like random forest significantly outperformed other investigated classifiers, including the commonly used multinomial logit model (MNL).

The features used in the application of this ML approach include origin, destination, day of the week, time of the day (in 1-hour segments), load size, and priority. These features provide a comprehensive view of the factors that influence transportation mode selection. While using such a large number of features can make the problem very complex when approached with a pure optimization strategy, it is relatively straightforward for a good classification algorithm like random forest. This algorithm can handle the complexity by evaluating the importance of each feature and making predictions based on the combined information.

By utilizing ML for transportation mode selection, we can achieve a higher level of accuracy and efficiency in determining the best transportation modes. This approach not only streamlines the decision-making process but also ensures that each load is transported in the most cost-effective and timely manner. The integration of ML techniques into transportation logistics represents a significant advancement in optimizing operations and reducing costs.

4.14 Random Forest for Transportation Mode Selection

Random forest is frequently utilized for transportation mode selection due to its robust features that make it ideal for such complex problems. One of the key advantages of random forest is its ability to handle a large number of variables effectively. Random forest excels in handling a multitude of variables and their interactions, which is crucial in transportation logistics [1]. This capability is particularly beneficial in multimode transportation, where multiple factors such as origin, destination, load size, and priority need to be considered simultaneously. The algorithm can process both two-class and multi-class classification problems, making it versatile for various transportation scenarios (Fig. 4.1).

Another significant benefit of random forest is its strong predictive performance, even in the presence of noisy data [4]. This resilience means that it does not require a rigorous preselection of features, which can be a time-consuming process. The ensemble nature of the random forest algorithm helps in avoiding overfitting, ensuring that the model generalizes well to unseen data. This feature is crucial for maintaining accuracy and reliability in predicting transportation modes under different conditions.

Random forest also excels in handling a combination of categorical and continuous predictors, making it suitable for the diverse nature of transportation data [8]. It tolerates interactions among predictor variables, allowing for a more comprehensive analysis of how different factors influence the mode selection. Additionally, high-quality and free implementations of random forest are readily available, such as

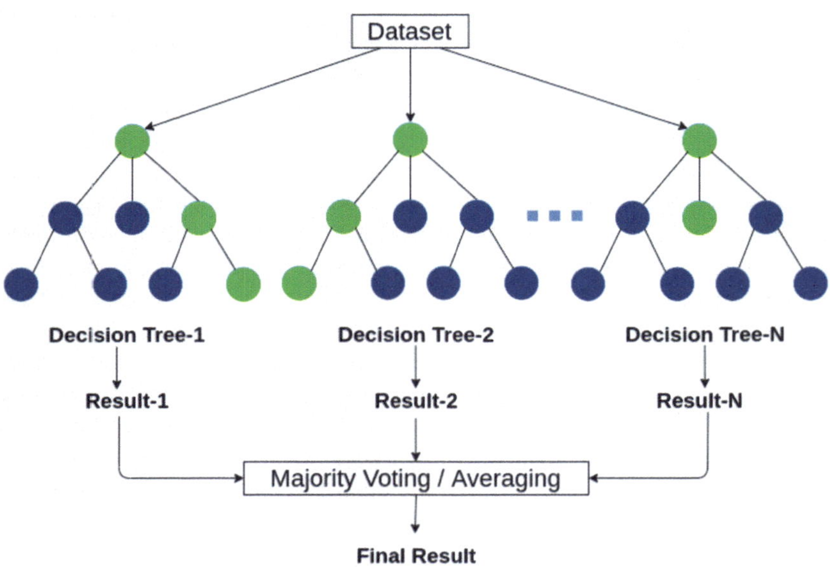

Fig. 4.1 Random forest for transportation

in the scikit-learn library for Python [15]. This accessibility enables practitioners to easily implement and experiment with random forest algorithms, enhancing the decision-making process in transportation logistics.

Overall, the random forest algorithm's flexibility, robustness, and ease of use make it an excellent choice for addressing the complexities of transportation mode selection. Its ability to manage large datasets with numerous variables and its resistance to overfitting contribute to its effectiveness in optimizing transportation operations.

4.14.1 Random Forest Parameters

In the implementation of a random forest algorithm for transportation mode selection, careful consideration of its parameters is essential to optimize performance and avoid common pitfalls such as overfitting and underfitting. One of the foundational steps in setting up the model is data splitting, where the dataset is divided into training and testing sets. For our purposes, a 70%/30% split is used. This means 70% of the data is used to train the model, allowing it to learn the underlying patterns, while the remaining 30% is reserved for testing the model's performance on unseen data.

The number of decision trees in the forest, denoted by the parameter $n\_estimators$, is set to 5000. Having a large number of trees helps improve the robustness and accuracy of the model, as it relies on the aggregate results of numerous decision trees to make predictions. However, it is also crucial to control the depth of each tree. In this case, the tree depth is limited to 10 to prevent overfitting. When trees grow too deep, they tend to capture noise and specific patterns in the training data that do not generalize well to new data.

Another important parameter is the minimum number of samples required to split an internal node, set at 20% of the total samples. Setting this parameter helps in controlling the tree's growth. If the value is too large, the model might underfit by failing to capture important patterns in the data. Conversely, if the value is too small, it might lead to overfitting. Similarly, the maximum number of leaf nodes is also set to 20%, balancing the complexity of the trees to avoid underfitting.

The criterion used to split nodes within the trees is the Gini impurity measure. The Gini criterion is chosen because it often provides better performance in classification tasks compared to alternatives like entropy. It measures the impurity or purity of a node, guiding the algorithm in making splits that increase the homogeneity of the resulting nodes, which helps in creating more accurate and reliable models.

4.14.2 Results

To evaluate the effectiveness of the machine learning model for transportation mode selection, we employed the area under the curve (AUC) metric, which provides a measure of the model's ability to distinguish between different classes. The AUC value obtained was around 0.85, indicating a strong performance in this application. An AUC of 0.85 suggests that the model has a high true positive rate and a low false positive rate, which is a satisfactory outcome for predicting transportation modes.

Additionally, the model achieved a recall rate of 92%, meaning that it correctly identified 92% of the relevant instances. This high recall rate is particularly advantageous in practical applications, as it ensures that most of the true transportation modes are captured by the model. This performance surpasses existing ad hoc methods, demonstrating the efficacy of the machine learning approach in handling the complexity and variability inherent in transportation logistics.

The algorithm's ability to learn from historical data and accurately forecast the necessary combination of transportation methods highlights its potential for improving operational efficiency. By leveraging past data, the model can predict transportation modes with high accuracy, providing a reliable basis for decision-making. This predictive capability is invaluable in planning and optimizing transportation strategies, ultimately leading to cost savings and enhanced service quality.

Moreover, the initial solutions generated by the machine learning model serve as an excellent starting point for further optimization. While the model's predictions are robust, they can be fine-tuned and enhanced through additional optimization algorithms tailored to specific scenarios and constraints. This iterative improvement process ensures that the transportation strategies are not only accurate but also optimized for performance and cost-efficiency, providing a comprehensive solution to the transportation mode selection problem.

References

1. L. Breiman, Random forests. Mach. Learn. **45**(1), 5–32, 2001.
2. K. Button, D.A. Hensher, *Handbook of Transport and the Environment* (Emerald Group Publishing Limited, London, 2001)
3. M. Christopher, M. Holweg, *Supply Chain 2.0: Managing Supply Chains in the E ra of Turbulence* (Springer, Berlin, 2011)
4. D.R. Cutler, T.C. Edwards Jr, K.H. Beard, A. Cutler, K.T. Hess, J. Gibson, J.J. Lawler, Random forests for classification in ecology. Ecology **88**(11), 2783–2792 (2007)
5. D.E. Goldberg, *Genetic Algorithms in Search, Optimization, and Machine Learning* (Addison-Wesley Longman Publishing Co., Inc., New York, 1989)
6. M.M. Golias, M.G. Karlaftis, I.E. Nikolaou, Logistical planning and optimization in transportation systems. Transp. Res. Part C Emerging Technol. **22**, 1–14 (2012)
7. J. Hagenauer, M. Helbich, A comparative study of machine learning classifiers for modeling travel mode choice. Expert Syst. Appl. **78**, 273–282 (2017)
8. J. Han, M. Kamber, J. Pei, Data mining applications in logistics and supply chain management. Comput. Ind. Eng. **57**(4), 1347–1364 (2009)

9. J.S.L. Lam, T. Notteboom, D. Yang, Q. Meng, Last-mile logistics: A review of challenges and opportunities. Transp. Rev. **38**(1), 24–47 (2018)
10. A. McKinnon, M. Browne, A. Whiteing, M. Piecyk, *Green Logistics: Improving the Environmental Sustainability of Logistics* (Kogan Page Publishers, London, 2014)
11. J. Monios, *Intermodal Freight Transport and Logistics* (Kogan Page Publishers, London, 2013)
12. T. Notteboom, J.-P. Rodrigue, Containerisation, box logistics and global supply chains: the integration of ports and liner shipping networks. Marit. Econ. Logist. **11**(1), 1–39 (2009)
13. T. Notteboom, B. Vernimmen, Container logistics in ports. Marit. Econ. Logist. **11**(3), 197–218 (2009)
14. I.H. Osman, J.P. Kelly, Metaheuristics: A bibliography. Ann. Oper. Res. **63**(5), 511–623 (1996)
15. F. Pedregosa, G. Varoquaux, A. Gramfort, V. Michel, B. Thirion, O. Grisel, M. Blondel, P. Prettenhofer, R. Weiss, V. Dubourg, et al., Scikit-learn: machine learning in python. J. Mach. Learn. Res. **12**, 2825–2830 (2011)
16. C. Prins, A review of metaheuristics for vehicle routing. J. Heuristics **10**(6), 587–614 (2004)
17. G.R. Raidl, J. Gottlieb, Metaheuristic optimization via memory and evolution: tabu search and scatter search. Handb. Metaheuristics **57**, 57–89 (2000)
18. D. Rodriguez, T. Tulla, J. Romera, Rail transportation and sustainability. Transp. Rev. **35**(5), 615–638 (2015)

Chapter 5
Optimization Methods for Multicast Routing Problems

Carlos A. S. Oliveira

Abstract In multicast networks, the objective is to send information to multiple users with a single transmission. This allows for savings in bandwidth since data can be shared across network links. An important problem in multicast networks is the determination of an optimum route to be followed by packets in a multicast group. This is known as the multicast routing problem (MRP). A large number of heuristic algorithms have been proposed in recent years to solve the MRP, which is of great interest to network engineers. In this paper, we propose a hybrid metaheuristic approach for the MRP, where a GRASP is used in conjunction with a variable neighborhood search algorithm to find near-optimal solutions to the MRP. Simulation results show that the proposed technique is superior in terms of solution quality.

Keywords Optimization · Routing · Multicast · Telecommunications · Machine learning

5.1 Introduction

Multicast services have become a cornerstone in the landscape of modern networked applications, enabling efficient and direct communication between a single source node and a specific set of destination nodes. This functionality is particularly valuable in scenarios where the same data must be delivered to multiple recipients simultaneously, such as in live video streaming, online gaming, and real-time data distribution in financial services.

With the rapid advancements in Internet and intranet technologies, particularly within corporate environments, the utilization of multicast services has seen a

C. A. S. Oliveira (✉)
Principal Inventive Scientist, Network Chief Technology Organization, AT&T Labs Inc., Bellevue, WA, USA
e-mail: carlos.oliveira@att.com

C. A. S. Oliveira, M. P. Pardalos (eds.), *Handbook of Artificial Intelligence and Data Sciences for Routing Problems*, Springer Optimization and Its Applications 219, https://doi.org/10.1007/978-3-031-78262-6_5

steady rise. These services offer significant advantages over traditional unicast transmission, where data is sent individually to each destination. By using multicast, network bandwidth is conserved, and latency is reduced, making it a preferred method for scenarios requiring simultaneous data delivery to multiple endpoints.

However, the wide-scale deployment of multicast applications faces several algorithmic challenges that have yet to be fully resolved. One of the most pressing issues is routing within multicast systems. Unlike traditional unicast systems, where the routing problem can be efficiently solved using algorithms such as Dijkstra's algorithm, multicast routing is more complex and is better represented by the Steiner tree problem. The Steiner tree problem is a well-known NP-hard problem, meaning that it is computationally infeasible to find the optimal solution for large instances within a reasonable time frame [4].

The inherent complexity of the multicast routing problem has led researchers to explore heuristic algorithms that provide good, albeit non-optimal, solutions. These heuristic methods include algorithms proposed by various researchers [2, 6, 7, 10–12], which attempt to approximate the optimal routing solution in a fraction of the time it would take to solve the problem exactly. However, these approaches typically lack any guarantee of local optimality, a significant drawback that becomes increasingly problematic as the size of the multicast group grows.

Efficient resource utilization is crucial in multicast systems, particularly in scenarios involving a large number of group members. The absence of local optimality guarantees in heuristic solutions can lead to suboptimal resource allocation, where network bandwidth and computational resources are not used as efficiently as possible. This inefficiency can be detrimental in large-scale applications, where even minor improvements in resource utilization can translate into substantial cost savings and performance enhancements.

To address these challenges, ongoing research in the field of multicast routing is focused on developing more sophisticated heuristic and approximation algorithms that strike a better balance between computational efficiency and solution quality. These algorithms aim to provide better guarantees of performance, particularly in terms of local optimality, while remaining feasible for large-scale implementations.

5.1.1 The Multicast Routing Problem

The primary goal of a multicast network is to facilitate efficient communication from a single source node to multiple destination nodes through a single send operation. This efficiency is achieved by strategically re-transmitting data whenever two or more destinations can be reached from a single intermediate node. This mechanism significantly reduces the overall network load compared to unicast transmissions, where data must be sent separately to each destination. In a multicast network, nodes that are interested in receiving the same piece of data are collectively referred to as a multicast group.

The key challenge in operating a multicast network lies in determining the optimal routes for delivering data to all members of a multicast group. Unlike unicast routing, which involves a straightforward path from the source to a single destination, multicast routing requires the creation of a more complex structure that connects the source to multiple destinations. The most common method for linking the source to its destinations is by constructing a tree that spans all the involved nodes, including both the source and the destinations. This structure, known as a multicast tree, ensures that data is transmitted efficiently across the network with minimal redundancy.

The specific objectives and constraints associated with constructing a multicast tree can vary significantly depending on the application and its performance requirements. For instance, in real-time video streaming applications, where timely delivery is crucial, quality of service (QoS) constraints often play a pivotal role. A common QoS requirement is that the transmission delay between the source and each destination must be below a fixed threshold, denoted as Δ. This constraint is critical in ensuring that video streams are delivered with minimal latency, maintaining the quality of the user experience.

Given this context, a fundamental objective in multicast routing is to minimize the total cost of the multicast tree while adhering to the QoS constraints. The cost of the tree typically reflects the use of network resources, such as bandwidth, and minimizing it helps to optimize the overall efficiency of the network. However, the requirement that all source-destination pairs have a delay less than Δ adds a layer of complexity to the problem. This specific routing problem, where the goal is to minimize the tree cost subject to delay constraints, is known as the delay-constrained multicast routing problem (DCMRP).

The DCMRP is particularly challenging due to its combinatorial nature and the conflicting objectives of minimizing cost while satisfying strict delay requirements. As the size of the multicast group and the network's complexity increase, finding an optimal solution becomes computationally infeasible, leading to the development of heuristic and approximation algorithms that aim to find near-optimal solutions within a reasonable time frame. These algorithms must carefully balance the trade-offs between cost and delay, often requiring sophisticated techniques to navigate the complex solution space.

5.1.2 Formal Statement

To provide a formal description of the DCMRP, let $G = (V, E)$ be a graph, where V is the set of nodes and E is the set of links between nodes. The source node is represented by s, and the destinations are $D = \{d_1, \ldots, d_k\}$, with $D \subset V$. There is a cost function $c : E \to \mathbb{Z}_+$ representing link costs and a delay function $\tau : E \to \mathbb{Z}_+$, giving the time elapsed when traversing an edge $e \in E$. The problem asks for a set of edges $E' \subseteq E$ such that s is connected to every node $d \in D$ on $G' = (V, E')$, and the maximum delay is bounded, i.e.,

$$\max_{d \in D} \sum_{e \in \mathcal{P}_d} \tau(e) \leq \Delta,$$

where $\mathcal{P}_d$ is the path in E' from s to d, such that the total cost $\sum_{e \in E'} c(e)$ of E' is minimum.

The DCMRP is easily seen as a generalization of the Steiner problem on graphs. In the Steiner problem, one is given a graph $G = (V, E)$ together with a cost function $c : E \to \mathbb{Z}_+$, and a set $R \subset V$ of required nodes. The nodes in $V \setminus R$ are called Steiner nodes. The objective is to find a tree T linking the nodes in D, passing through Steiner nodes if necessary, such that the cost $\sum_{e \in T} c(e)$ is minimum. The Steiner problem on graphs is known to be NP-hard [4].

Theorem 1 *Theorem: The DCMRP is NP-hard.*

Proof Consider the following transformation between instances of the Steiner problem to instances of the DCMRP. Given an instance of the Steiner tree problem, construct an instance of DCMRP with the same graph. Select a node among the required nodes to become the source and let the remaining required nodes be destinations. Set Δ to infinity. An optimal solution to the transformed problem will be a solution to the original instance of the Steiner problem. Similarly, an optimal solution for the original instance will give an optimal solution for the transformed instance. Thus, we conclude that the DCMRP is NP-hard. □

Given the computational complexity of the DCMRP, our interest is in finding efficient algorithms that return good, near-optimal solutions for a large number of practical instances.

5.1.3 Math Programming Formulation

The delay-constrained multicast routing problem (DCMRP) can be rigorously defined using a mixed-integer linear programming (MILP) model. This formulation considers a graph $G = (V, E)$, where V denotes the set of nodes and E represents the set of edges (links) between these nodes. The objective is to determine the minimum-cost multicast tree that connects a source node $s \in V$ to a set of destination nodes $D = \{d_1, d_2, \ldots, d_k\} \subset V$, while ensuring that the transmission delay from the source to any destination does not exceed a specified threshold Δ.

We are given the following:

- Graph: $G = (V, E)$.
- Source node: $s \in V$.
- Destination nodes: $D = \{d_1, d_2, \ldots, d_k\} \subset V$.
- Cost function: $c : E \to \mathbb{Z}_+$ where $c(e)$ represents the cost associated with edge $e \in E$.

- Delay function: $\tau : E \to \mathbb{Z}_+$ where $\tau(e)$ represents the delay associated with edge $e \in E$.
- Delay bound Δ, which specifies the maximum allowable delay from the source s to any destination $d \in D$.

The goal is to find a subset of edges $E' \subseteq E$ such that the subgraph $G' = (V, E')$ forms a tree connecting the source node s to all destination nodes $d \in D$. Additionally, the tree should minimize the total cost $\sum_{e \in E'} c(e)$, subject to the constraint that the delay from s to each destination node d does not exceed Δ.

We introduce the following decision variables:

- Binary variables:

$$x_e = \begin{cases} 1 & \text{if edge } e \text{ is included in the multicast tree,} \\ 0 & \text{otherwise,} \end{cases}$$

for each $e \in E$.
- Continuous variables:

$$y_{i,j}^d \geq 0,$$

representing the flow from node i to node j towards destination d, for each $(i, j) \in E$ and $d \in D$.

The objective is to minimize the total cost of the multicast tree:

$$\text{Minimize} \sum_{e \in E} c(e) \cdot x_e.$$

The problem is subject to the following constraints:
Flow Conservation Constraints:

$$\sum_{j:(i,j) \in E} y_{i,j}^d - \sum_{j:(j,i) \in E} y_{j,i}^d = \begin{cases} |D| & \text{if } i = s, \\ -1 & \text{if } i = d, \\ 0 & \text{otherwise (i.e., for } i \neq s \text{ and } i \neq d), \end{cases}$$

for all $i \in V$ and $d \in D$. This set of constraints ensures that flow is properly routed from the source s to each destination d, maintaining flow balance at all intermediate nodes.

Edge Capacity Constraints:

$$y_{i,j}^d \leq |D| \cdot x_{(i,j)}, \quad \forall (i, j) \in E, \forall d \in D.$$

These constraints ensure that flow is only routed through edges that are included in the multicast tree.

Delay Constraints:

$$\sum_{(i,j)\in\mathcal{P}_d} \tau(i,j)\cdot x_{(i,j)} \leq \Delta, \quad \forall d \in D.$$

This constraint guarantees that the delay from the source s to each destination d does not exceed the specified threshold Δ.

Subtour Elimination Constraints (Optional): Additional constraints can be introduced to prevent cycles in the multicast tree, ensuring that the selected edges form a tree structure. This can be achieved using either subtour elimination constraints or flow-based formulations, depending on the preferred approach.

The MILP formulation provided above captures the essence of the DCMRP by balancing cost minimization with delay constraints across multiple destinations. Given the combinatorial nature of the problem, which is related to the NP-hard Steiner tree problem, solving the DCMRP for large instances may require the use of specialized solvers or heuristic methods that can approximate the optimal solution within reasonable time limits. This formulation serves as a robust foundation for further exploration and solution methodologies aimed at improving the efficiency and reliability of multicast routing in delay-sensitive applications.

5.1.4 Overview

In this chapter, we are interested in solution methods for the delay-constrained multicast routing problem. In particular, we propose a new method for computing routing trees for multicast networks using a hybridization of GRASP and variable neighborhood search. The strategy is employed to avoid suboptimal solutions produced by existing heuristic algorithms.

At the same time, we propose an innovative way of working with the GRASP metaheuristic. The new method combines elements of the variable neighborhood search algorithms in a way that produces fast implementations for the problem considered. The paper is organized as follows: in the next section, we describe the basic steps of the metaheuristic. Next, we propose modifications to the GRASP framework that can be used to improve performance. Next, we discuss computational results. In the last section, we provide some concluding remarks.

5.2 Algorithm Description

The development of metaheuristics has started from the realization that practical algorithms are necessary to cope with the high computational complexity of some important problems. Most of these techniques are based on finding solutions close to the global optimum with the help of gradient methods combined with randomization

Algorithm 4 GRASP

read instance data
initialize data structures
while termination criterion not satisfied **do**
    s ←create solution
    improve solution s
end while

rules to avoid local optima. Therefore, most metaheuristics use similar principles in different ways. A conceivable goal for the algorithm designer is, therefore, to borrow techniques from different metaheuristics to create algorithms that reflect characteristics of the problem at hand. In this paper, we use some of the techniques in the variable neighborhood search metaheuristic in a GRASP implementation to solve the DCMRP.

In constructing hybrid metaheuristics, we acknowledge that the main contribution of metaheuristic development is not the creation of new algorithms but rather the development of intelligent strategies for mixing existing non-optimal techniques. The existing non-optimal algorithms are frequently dependent on the problem being solved. However, high-level patterns of algorithm combination can emerge and be reused as necessary.

5.2.1 Using GRASP

The greedy randomized adaptive search algorithm (GRASP) is a metaheuristic proposed by Feo and Resende [3] aimed at finding near-optimal solutions for combinatorial optimization problems. It is composed of a number of iterations, where a new solution is picked from the feasible set using a construction algorithm and subsequently improved using some local search method. GRASP has been very successful in a number of applications such as QAP [8], frequency assignment [5], and satisfiability, occurring in diverse fields. The steps of GRASP are summarized in Algorithm 4.

The GRASP algorithm is a multistart method, where at each iteration, a new solution is constructed and subsequently improved. The algorithm presented in 4 can be described as a general iterative optimization framework, which is commonly used in heuristic and metaheuristic approaches to solving complex optimization problems. Below is a detailed explanation of each step involved in the algorithm.

The algorithm begins by reading the instance data, which involves loading the specific problem parameters and structures required for the optimization process. This step is crucial as it sets up the environment with all necessary information, such as the graph structure, cost and delay functions, and any constraints specific to the problem being addressed. This data provides the foundation upon which solutions will be generated and evaluated.

Following this, the algorithm proceeds to initialize data structures. This step involves setting up any variables, arrays, or other data constructs that will be used throughout the optimization process. These data structures might include representations of potential solutions, tracking variables for performance metrics, or structures used to store intermediate results. Proper initialization is essential to ensure that the algorithm can operate efficiently and that memory is managed effectively during execution.

The core of the algorithm is encapsulated within a while loop, which continues to execute until a predefined termination criterion is met. This criterion could be based on a fixed number of iterations, a time limit, or a convergence condition where improvements to the solution become negligible. Within each iteration of the loop, the algorithm follows two main steps.

First, the algorithm creates a solution (denoted as s). This step involves generating an initial feasible solution to the problem based on the current state of the data and any heuristic rules or strategies defined within the algorithm. The method used to create this solution can vary significantly depending on the problem type and the specific heuristic or metaheuristic being applied. It could involve constructing a solution from scratch or making modifications to an existing solution.

After generating the initial solution, the algorithm moves to the improvement phase, where it attempts to enhance the quality of the solution s. This improvement step could involve local search techniques, where the algorithm explores the neighborhood of the current solution to find a better one, or more sophisticated optimization strategies, such as simulated annealing, genetic algorithms, or tabu search, which systematically explore the solution space. The objective during this phase is to refine the solution s to either reduce the cost, improve performance metrics, or better satisfy the problem constraints.

The loop continues, iteratively generating and improving solutions until the termination criterion is satisfied. Once the loop exits, the algorithm typically returns the best solution found during its execution, which is then analyzed or used as the final solution to the problem at hand.

This iterative approach allows the algorithm to gradually converge towards an optimal or near-optimal solution, making it well-suited for solving complex problems where exact methods may be computationally infeasible.

5.2.2 The Construction Phase

Algorithm 5 follows a greedy randomized construction approach, commonly used in metaheuristic techniques such as the greedy randomized adaptive search procedure (GRASP). This approach combines elements of both greedy algorithms and randomization to build a solution iteratively. Below is a detailed explanation of how the algorithm operates.

Algorithm 5 GRASP construction phase

```
while solution s is not complete do
    Sort list L of k existing candidate elements from best to worst
    Let α be a parameter from interval 0 < α ≤ k
    RCL ← set of α best candidates in L
    Randomly select from RCL with uniform probability
end while
```

The algorithm operates within a while loop that continues until the solution s is fully constructed. The notion of a complete solution depends on the specific problem being addressed; for example, in a routing problem, a solution might be considered complete when all required nodes have been visited. Each iteration of the loop focuses on incrementally building the solution by adding one element at a time.

At the start of each iteration, the algorithm sorts the existing candidate elements. These candidates represent potential additions to the current partial solution and are sorted based on a problem-specific criterion, typically related to cost, distance, or another objective function relevant to the optimization task. Sorting ensures that the algorithm can prioritize elements that are more likely to lead to an optimal solution.

After sorting, the algorithm uses an integer parameter α such that $0 < \alpha \leq k$, where k is the total number of candidate elements. The value of α determines the number of top candidates that will be included in the restricted candidate list (RCL). This list contains the α best candidates according to the sorting criterion. The concept of an RCL allows the algorithm to strike a balance between greediness and randomness: by only considering the top α candidates, the algorithm ensures that it is focusing on promising options while still maintaining some level of diversity in the search process.

Once the RCL is constructed, the algorithm **randomly selects an element** from this list with uniform probability. This random selection introduces a stochastic component to the algorithm, preventing it from always following a deterministic greedy path. By doing so, the algorithm explores different regions of the solution space across multiple runs, increasing the likelihood of finding a globally optimal or near-optimal solution.

This process of sorting, creating an RCL, and randomly selecting an element continues until the solution s is complete. By iteratively constructing the solution in this manner, the algorithm effectively combines the efficiency of greedy heuristics with the exploration capabilities of randomized algorithms. This approach is particularly useful in combinatorial optimization problems, where the solution space is vast, and traditional deterministic methods may struggle to find high-quality solutions within a reasonable time frame. The randomization aspect ensures that the algorithm can escape local optima, while the greedy component ensures that the constructed solution remains competitive.

Algorithm 6 Improved construction for GRASP

```
while solution s is not complete do
    Select a random α, such that 1/n ≤ α ≤ k
    k ← 1/α
    c ← ∞ /* if this is a minimization problem */
    for j = 1, ..., k − 1 do Let C_j be the list of candidates at this iteration
        Randomly select one element x in C_j
        c_j ← f(x)
        if c_j < c then c ← c_j
            y ← x
        end if
    Insert element y in the solution s
    end for
end while
```

.

5.2.3 Improving the Construction Phase

In this section we describe a new method for the GRASP construction phase that is on average equivalent to the existing method but much more efficient in practice. The method uses the following observation:

Observation 1 *Let* $x_1, \ldots, x_n$ *be an unordered sequence, and* $y_1, \ldots, y_n$ *the corresponding ordered sequence. Then, to find a random element among* $y_1, \ldots, y_\alpha$, *for* $0 < \alpha \le n$ *is on average equivalent to selecting the best of* 1α *random elements of* $x_1, \ldots, x_n$.

Proof Given the sequence $x_1, \ldots, x_n$, by selecting random elements of the sequence, the probability of selecting one of the elements in the RCL is $n\alpha/n = \alpha$. Consequently, if W is an indicator random variable defined as 1 whenever the element picked is in the RCL, then $E(W) = \alpha$. Then, after doing k independent trials, the average number of elements in the RCL is $k\alpha$, by the additivity of expectation. If we have $k = 1/\alpha$, then on average, there is just one element of the RCL within the picked elements. Now, to know which of these elements is the one in the RCL, we just need to take the smallest one. □

The observation above gives a very efficient way of implementing the RCL test, which gives, on average, the same results. Start with the full set C of candidate elements. Then, at each step, generate a value of α and pick at random $k = 1/\alpha$ elements of C. From the picked elements, store only the one that is the best fit for the greedy function. This method is depicted in Algorithm 6.

A clear advantage in terms of computational complexity is achieved by the proposed construction method for GRASP. The best advantage is that, while in the original technique, the candidate elements must be sorted, this is not necessary in the proposed algorithm. Moreover, the complexity of the traditional construction is dependent on the number of candidate elements. In our method, the complexity is constant for a fixed value of α. For example, if alpha is $n/2$, then we need just two iterations to find an element in the RCL, with high probability.

Theorem 2 *The complexity of selecting elements from the RCL in the modified construction algorithm is* $n \log n$.

Proof A probabilistic analysis of the complexity of the algorithm will be used. The important step that must be accounted for is the selection of a random element from C. Initially, let us assume that at each iteration of the for loop, the size of the list is n. Knowing that at each step the value of α is chosen from a uniform distribution, we find that the average number N of elements selected in the for loop is

$$
\begin{aligned}
E(N) = E(\sum_{j=1}^{n} k) &= \sum_{j=1}^{n} E(k|\alpha) \\
&= \sum_{j=1}^{n} \int_{1/n}^{1} E(k|\alpha = t)\, dt \\
&= \sum_{j=1}^{n} \int_{1/n}^{1} 1/t\, dt \\
&= \sum_{j=1}^{n} (\log 1 - \log 1/n) \\
&= \sum_{j=1}^{n} \log n = n \log n.
\end{aligned}
$$

Now, the decrease in the size of the list will not change the complexity of the result, since

$$
\sum_{j=1}^{n} \log(n - j) = n \log n + \log \pi + \log \frac{\csc n\pi}{(n-1)!}.
$$

The second term is a constant, and the third goes to zero as n tends to infinity very fast, because of the value $(n-1)!$ in the denominator. □

5.2.4 Path Relinking Phase

GRASP has the advantage of being easy to develop since it is composed of relatively independent procedures (the constructor and local search phases). It is well suited for applications with existing heuristic algorithms that can be combined with GRASP to find a better solution.

Algorithm 7 Path relinking for GRASP

```
Input: initial_solution (the initial solution to improve)
PartialSolutions (a list of parts of solutions to use for improvement)
Output: improved_solution (the improved solution)
improved_solution ← initial_solution
for each part in PartialSolutions do
    if CanBeInserted(improved_solution, part) then
        improved_solution ← InsertPart(improved_solution, part)
    end if
end for
return improved_solution
```

However, one of the weaknesses of GRASP is its incapacity to integrate good solutions found previously into the current search iteration. Since each iteration will create a completely different solution, there is no information added to the system when a good solution is found.

A method that has been used lately to overcome this problem is called *path relinking* (PR) [1, 9]. In PR, a subset of the best solutions found is kept in a separate memory called the *elite set*. At each iteration, one of the solutions s will be selected, and a process of comparing the current solution with s will start. Each component of the solution will be changed to the corresponding value on s, and after this, a local search will be initiated to check for local optimality.

The disadvantage of PR is the large time it takes to run, in comparison with the rest of the GRASP algorithm. This, in practice, has been a restricting factor in the use of PR on practical applications. Although PR brings a relative boost in solution quality on each iteration, the effect can be negative since the number of iterations may be reduced due to the added complexity.

We introduce a new method for taking advantage of existing solutions produced by GRASP. Instead of the strategy taken by PR, which stores complete solutions from the history of iterations, we store only parts of solutions that can be freely mixed into the improvement phase. The method used is described in Algorithm 7.

5.2.5 Evaluating the Algorithms

To solve the multicast routing problem (MRP), the goal is to identify a tree structure that effectively links the source node to multiple destination nodes while adhering to specified delay constraints. This problem can be viewed as a variant of the classic Steiner tree problem, where the additional requirement is that the paths from the source to each destination must not exceed a predetermined delay threshold. The Steiner tree problem itself is known to be NP-hard, making it computationally challenging, and the inclusion of delay constraints further complicates the search for an optimal solution. Therefore, effective heuristics and approximation algorithms are essential for tackling this problem, particularly in large-scale networks.

In this context, we explore the use of different algorithms during the construction phase of the greedy randomized adaptive search procedure (GRASP) to generate initial solutions for the MRP. The construction phase is crucial as it sets the foundation for the entire optimization process, and its efficiency can significantly impact the overall performance of the algorithm. Specifically, if the construction phase is too slow, it can become a bottleneck, hindering the algorithm's ability to quickly explore the solution space and refine the initial solutions through subsequent improvement phases. As a result, the choice of algorithm for this phase must strike a balance between solution quality and computational efficiency.

To assess the impact of different approaches in the construction phase, we experimented with two distinct implementations: the random tree method and the Kou-Markowsky-Berman (KMB) heuristic. The random tree method involves generating a tree by randomly selecting edges while ensuring that all nodes in the multicast group are connected. This approach is straightforward and computationally inexpensive, making it an attractive option for scenarios where speed is prioritized. However, the randomness of the method can result in suboptimal trees, particularly in terms of cost and delay, which may require significant refinement in later stages of the GRASP process.

On the other hand, the KMB heuristic is a more sophisticated approach specifically designed for Steiner tree problems. It constructs a tree by iteratively adding the shortest paths between nodes until all required nodes are connected. The KMB heuristic is known for its ability to produce high-quality solutions that are close to optimal, particularly for problems involving large numbers of nodes and complex network topologies. However, the trade-off is that it is more computationally intensive than the random tree method, which could potentially slow down the construction phase if not properly managed.

By comparing these two methods in the GRASP construction phase, we aim to evaluate how the choice of algorithm affects both the efficiency and effectiveness of the overall solution process. The random tree method offers speed and simplicity, which can be beneficial in time-sensitive applications or when dealing with extremely large networks. In contrast, the KMB heuristic may provide better initial solutions at the cost of increased computational effort, which could ultimately lead to more refined and cost-effective multicast trees after the improvement phase. Our investigation will help determine the most suitable approach for different types of MRP instances, balancing the need for quick construction with the desire for high-quality solutions that meet stringent delay constraints.

5.3 Concluding Remarks

The delay-constrained multicast routing problem (DCMRP) represents a significant challenge in network design, particularly in the context of multicast services where efficient data transmission to multiple destinations is crucial. The inherent complexity of the problem, which generalizes the NP-hard Steiner tree problem, underscores the need for sophisticated algorithms that can balance computational feasibility with solution quality.

Heuristic and metaheuristic approaches, such as the greedy randomized adaptive search procedure (GRASP), have been explored as viable strategies for tackling DCMRP. These methods provide near-optimal solutions within reasonable time frames, making them suitable for large-scale implementations where exact algorithms would be computationally prohibitive. The GRASP method, particularly when hybridized with techniques like variable neighborhood search (VNS), offers a flexible and powerful framework for navigating the complex solution space of DCMRP.

However, the development of these algorithms is not without challenges. Ensuring local optimality and integrating previously found good solutions into current iterations remain areas for improvement. Path relinking (PR) has been proposed as a solution to these issues, although its computational overhead may limit its practicality in certain applications.

References

1. R.M. Aiex, S. Binato, M.G.C. Resende, Parallel GRASP with path-relinking for job shop scheduling. Parallel Comput. **29**, 393–430 (2003)
2. A. Ballardie, P. Francis, J. Crowcroft, Core-based trees (CBT)—an architecture for scalable inter-domain multicast routing. Comput. Commun. Rev. **23**(4), 85–95 (1993)
3. T.A. Feo, M.G.C. Resende, Greedy randomized adaptive search procedures. J. Glob. Optim. **6**, 109–133 (1995)
4. M.R. Garey, D.S. Johnson, *Computers and Intractability: A Guide to the Theory of NP-completeness* (W.H. Freeman and Company, New York, 1979)
5. F.C. Gomes, P.M. Pardalos, C.A.S. Oliveira, M.G.C. Resende, Reactive GRASP with path relinking for channel assignment in mobile phone networks, in *Proceedings of the 5th International Workshop on Discrete Algorithms and Methods for Mobile Computing and Communications* (ACM Press, New York, 2001), pp. 60–67
6. S. Hong, H. Lee, B.H. Park, An efficient multicast routing algorithm for delay-sensitive applications with dynamic membership, in *Proceedings of IEEE INFOCOM'98* (1998), pp. 1433–1440
7. V.P. Kompella, J.C. Pasquale, G.C. Polyzos, Optimal multicast routing with quality of service constraints. J. Netw. Syst. Manag. **4**(2), 107–131 (1996)
8. C.A.S. Oliveira, P.M. Pardalos, M.G.C. Resende, GRASP with path-relinking for the QAP, in *5th Metaheuristics International Conference*, Kyoto, Japan (2003), pp. 57.1–57.6
9. M.G.C. Resende, C.C. Ribeiro, A GRASP with path-relinking for private virtual circuit routing. Networks **41**, 104–114 (2003)
10. H. Salama, D. Reeves, Y. Viniotis, A distributed algorithm for delay-constrained unicast routing, in *Proceedings of the IEEE INFOCOM'97*, Kobe, Japan (1997)
11. R. Sriram, G. Manimaran, C. Siva Ram Murthy, A rearrangeable algorithm for the construction of delay-constrained dynamic multicast trees. IEEE/ACM Trans. Netw. **7**(4), 514–529 (1999)
12. Q. Zhu, M. Parsa, J.J. Garcia-Luna-Aceves, A source-based algorithm for delay-constrained minimum-cost multicasting, in *Proceedings of the IEEE INFOCOM95* (1995), pp. 377–385

Chapter 6
An Introduction to AI and Routing Problems in Mobile Telephony

Carlos A. S. Oliveira

Abstract Routing optimization in mobile telephony aims to enhance network efficiency and service quality by addressing complex and dynamic challenges such as call routing, handover management, load balancing, and energy consumption. Artificial intelligence (AI) and machine learning techniques are key to solving these problems, allowing for real-time, data-driven decisions that optimize resources, reduce latency, and ensure seamless communication. This paper explores various aspects of routing optimization, including cell site placement, radio resource allocation, quality of service (QoS)-based routing, and network security, highlighting AI's crucial role in modern mobile networks.

Keywords Optimization · Routing · Telephony · Telecommunications · Interference

6.1 Introduction

Routing optimization problems in mobile telephony pertain to improving the efficiency and quality of service in mobile communication networks. These problems can be complex and dynamic due to the mobility of users and the constantly changing network conditions.

Several problems arise when solving the general task of routing optimization in mobile telephony, including the following:

- Call Routing Optimization: The problem consists of efficiently routing voice calls or data traffic to their destinations while minimizing latency, congestion, and costs. AI-driven algorithms and network optimization techniques can be used

C. A. S. Oliveira (✉)
Principal Inventive Scientist, Network Chief Technology Organization, AT&T Labs Inc., Bellevue, WA, USA
e-mail: carlos.oliveira@att.com

C. A. S. Oliveira, M. P. Pardalos (eds.), *Handbook of Artificial Intelligence and Data Sciences for Routing Problems*, Springer Optimization and Its Applications 219, https://doi.org/10.1007/978-3-031-78262-6_6

to optimize call routing based on real-time network conditions, traffic patterns, and service quality requirements.

- Handover and Roaming Optimization: The issue entails ensuring seamless handovers between cell towers or across different networks (e.g., when roaming) to prevent call drops and maintain quality. AI algorithms have been developed to predict handover events, evaluate available networks, and manage handovers to minimize disruption and maintain call quality.
- Load Balancing: This is the problem of distributing mobile network traffic evenly across base stations to prevent congestion, reduce call drops, and optimize network utilization. A possible solution to this problem is to use machine learning models that can predict traffic patterns and dynamically allocate users to different base stations based on load and resource availability.
- Cell Site Placement and Coverage Optimization: Here, the goal is to determine optimal locations for cell towers and adjusting their coverage areas to ensure efficient network coverage and minimal interference. A solution can be proposed using techniques such as genetic algorithms, reinforcement learning, and optimization models that can help optimize cell site placement and coverage patterns.
- Optimizing Radio Resource Allocation: This happens when allocating radio resources, such as frequency channels and time slots, to users and services to maximize spectral efficiency and minimize interference. In this case, AI can dynamically allocate radio resources based on user demands, channel quality, and network conditions.
- QoS-Based Routing: This problem consist of routing mobile traffic to ensure quality of service (QoS) requirements are met for different applications (e.g., voice, video, data) with varying QoS demands. A possible solution strategy involves AI-driven routing algorithms that prioritize traffic based on QoS requirements, taking into account application-specific needs.
- Optimizing Backhaul Links: This is the problem of efficiently managing the links that connect cell towers to the core network to prevent bottlenecks and ensure reliable service. Using AI, one can optimize backhaul link routing, taking into account network conditions and traffic demands.
- Network Slicing for 5G: In 5G networks, the dynamic creation of network slices for different services and applications requires efficient routing and resource allocation. In this situation, AI algorithms can be used to automate the creation, management, and optimization of network slices, ensuring they meet service requirements.
- Energy-Efficient Mobile Networks: In this case, the problem is reducing the energy consumption of base stations and mobile devices while maintaining network performance. AI-driven solution techniques can optimize the power management of base stations and mobile devices, considering factors like traffic load and coverage.
- Network Security Routing: Finally, this problem in network routing consists of protecting mobile networks from threats and attacks while ensuring network availability and service quality. A solution based on AI can detect security threats in mobile traffic and reroute traffic to prevent or mitigate attacks. Routing

optimization problems in mobile telephony often require real-time decision-making due to the dynamic nature of wireless communication. AI and machine learning are instrumental in addressing these challenges by making adaptive and data-driven routing decisions to improve network efficiency and quality of service.

In this chapter we discuss several concepts related to optimization and routing on mobile telephony. The themes considered in this chapter include the following: First, some examples of problems occurring in mobile telephony will be presented. Section 6.2 provides a brief summary of the history and development of mobile telephony systems. The next section introduces some technologies in mobile telephony, along with its main characteristics. Finally, the next next section discusses many of the concepts used in this chapter, along with solution strategies.

6.2 Mobile Telephony Concepts

The history of the use of electromagnetic waves for data transmission had its beginnings with the studies of the German physicist Heinrich R. Hertz (1857–1894). Based on the physical properties predicted by electromagnetic theory, developed by James Clerk Maxwell (1831–1879), Hertz was the first researcher to create electromagnetic waves in an artificial way. The first services to use the propagation of electromagnetic waves were the beginnings of the telegraph system. Starting from there, the services of radio, television, maritime communication, radio amateur, army communication systems, and satellites, among others, have made use of the frequency spectrum in order to perform data transmission.

In order to improve the rational use of the frequency spectrum, the international community decided to create the ITU (International Telecommunication Unit) in 1932. This entity, derived from the old International Telegraph Union founded in 1865, was subsequently incorporated in 1947 as an agency of the United Nations. Its main objective is to cooperate with all nations and member states on allocating frequencies to specific services. This allows the creation of international standards, avoiding that different countries create incompatible legislation. Without an international standard, great confusion could ensue, since the use of the frequency spectrum is not easily restricted to geographic frontiers. As part of its mission, the ITU defines several services that further restrict the use of the frequency spectrum. Some of the services defined by ITU are listed on Table 6.1.

According to the ITU, cellular telephony is classified as a terrestrial mobile service. This classification encompasses all communications performed on land by mobile devices. Other channels of the spectrum are allocated to marine and aerial communication, with an additional list that extends to all other established services, including radio-amateur and analog and digital TV transmission, among others.

In its first iteration, which started roughly at the end of the 1970s, mobile telephony was introduced as a communication service exclusive to the use of

Table 6.1 Some of the services recognized by ITU

Fixed aeronautics	Mobile aeronautics
Aeronautical mobile satellite	Aeronautical radio navigation
Aeronautical radio navigation—satellite	Radio amateur
Radio amateur—satellite	Radio and TV transmission
Radio and TV transmission—satellite	Terrestrial mobile
Mobile satellite terrestrial	Meteorologic satellite
Radio astronomy	Radio navigation
Special operations	Standard time

automobiles. The main reason, other than the high cost of installation, was that available devices at the time depended on auto batteries in order to work. Energy storage technology still needed to go through many advances before it was possible to operate a cellular phone with its own battery. That first generation of mobile phones used an analog transmission technology, known as AMPS (advanced mobile phone system).

The end of the 1980s, however, witnessed a big jump on the availability of cellular phone systems, with the development of new devices with independent batteries. This gave rise to the development of even smaller phone sets, with greater autonomy. The concurrent development of the GSM (Global System for Mobile Communications) also had positive consequences, allowing the creation of a new generation of digital technologies, which largely replaced the previous models.

Currently, the area of mobile cellular telephony is still in a phase of quick development, characterized by global competition of large corporations for the control of the market. This scenario has resulted in a large number of innovations both on the available services and in the capacity of the system. The current tendency of the market is towards integration between voice and data, creating a structure that interconnects mobile telephony and the Internet [6].

6.3 Fundamental Definitions

6.3.1 Cells and Base Stations

A cellular telephony network is composed of two main elements: reception antennas and mobile devices. The antennas are placed at strategic points, known as base stations, located within their respective cells.

One of the main characteristics of a cellular mobile system is that the service coverage area is divided into several smaller regions, called cells. The format and extension of cells are defined based both on the geography of the desired region and the characteristics of the technology that is employed. One of the simplest ways to divide service regions is the construction of hexagonal cells. However, other forms

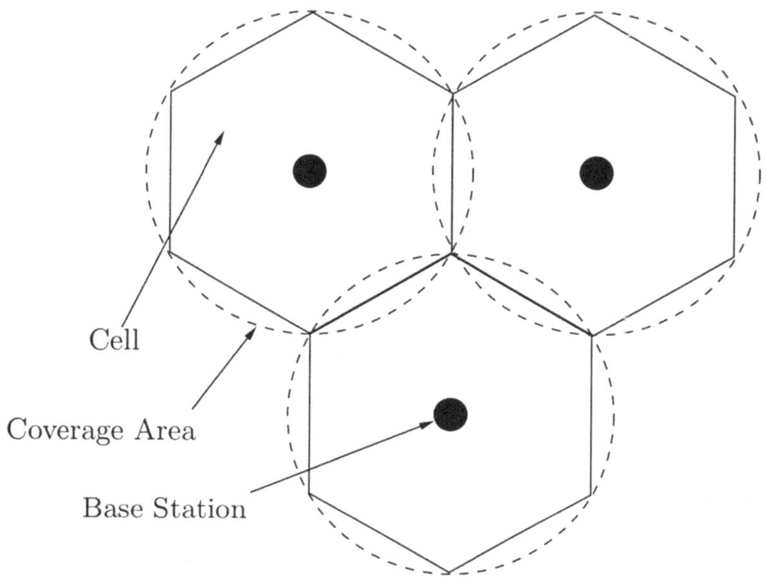

Fig. 6.1 Representation of a region divided into cells

of dividing the service area have also been adopted. For example, in cities with tall buildings, a division with cells corresponding to blocks may be more effective, due to the rectangular way those buildings are organized and the resulting user concentration. For another common example, a simple linear arrangement of cells is effective enough for the coverage of highways.

One can alternatively characterize a cell as the area covered by a single base station. In Fig. 6.1 one have a schematic representation of cell, with their respective coverage areas. Notice that there is an intersection between the areas covered by two adjacent cells. These intersections will be an important concern when we consider interference in the assigned frequency spectrum.

6.3.2 Frequency Spectrum

Radio frequency waves are an integral part of the spectrum of electromagnetic waves. The electromagnetic spectrum range goes from low-frequency waves to X-rays, including the range of visible light. Radio waves form the region of the spectrum commonly reserved for telecommunications.

A range of spectrum dedicated to a particular use is called a band. For example, in the United Kingdom, the range of spectrum dedicated to mobile telephony goes from 890 to 915 Hz, and from 935 to 960 Hz [7]. This band of spectrum is further divided into smaller regions, called frequency channels (or simply channels when

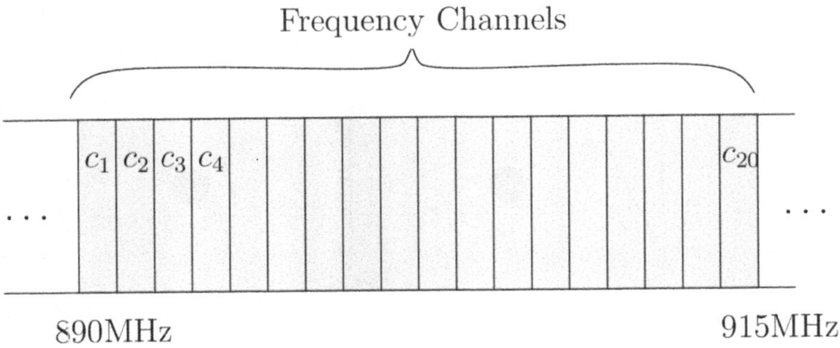

Fig. 6.2 Frequency spectrum and sample region located between frequencies 890 MHz and 915 MHz. The region is divided into channels $c_1, c_2, \ldots, c_{20}$

the meaning is made clear from the context). A channel can be identified by its central frequency. Therefore, it is common to read references in the literature to a "channel of 600 Hz," when in fact one is referring to the channel in which that is the central frequency (see Fig. 6.2).

As the number of available channels is limited, one can assign to each channel a unique integer number $c_i \in N$. This way, one can easily refer to the spectrum of available channels as a set $C \subset N = \{c_1, \ldots, c_n\}$. This way, when allocating channels to antennas, it is not necessary to refer directly to channels, but instead to integer numbers. This simplifies considerably the formalization of a model for the frequency assignment problem, as shown in the following sections.

In a mobile telephony system, each base station is responsible for the handling of users located in the region covered by their corresponding cells. As the number of users (user density) of each region is different, base stations are also subject to varying demand.

The demand at a base station is defined as the number of frequency channels necessary to service (or cover) users located at a mobile cell. The more users need to be serviced by a cell, the bigger the demand for frequency channels.

6.3.3 *Types of Interference*

Interference is a phenomenon occurring in all systems that transmit data using radio waves. It exists whenever a receiving device is exposed to radio transmissions from two sources that operate in the same, or nearly the same, frequencies (see Fig. 6.3). In general, the occurrence of interference makes it harder to provide adequate communication between devices due to the undesired noise in the transmission. Therefore, a well-known measure of quality for radio transmissions is the interference level occurring between components of the system. One must notice that interference can be measured precisely only on the receiving end, and therefore, it is a function of reception capacity.

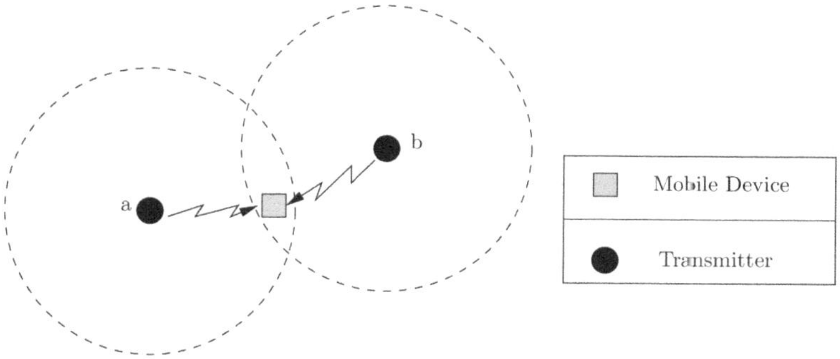

Fig. 6.3 Transmitters *a* and *b*, using the same channel, cause interference in a mobile device located in the shared region of two cells

On mobile telephony, the interference problem happens mainly due to the limits imposed by the available spectrum. For example, when the number of antennas is greater than the number of channels (something that happens in many cases), then a few of those channels will be needed to be used in more than one antenna. However, it is necessary that this kind of channel reuse be made in such a way that the resulting interference is still maintained at reasonable levels.

The types of channel interference are classified according to the proximity of transmitters. Co-site interference occurs among transmitters that are co-located, i.e., they share the same cell. Otherwise, the generated interference is referred to as far-site. Far-site interference, on the other hand, can be further classified according to the proximity of channels that generate the interference. Among the resulting type of far-site interference, the most important are *co-channel* and *adjacent-channel* interference.

6.3.3.1 Co-site Interference

This kind of interference occurs only among transmitters that are situated in the same geographical location. In the case of cellular telephony, co-site interference can happen between each pair of antennas operating in the same base station. Therefore, the possibility of co-channel interference increases along with the number of transmitters in a base station.

6.3.3.2 Far-Site Interference

This kind of interference occurs between any two transmitters that are geographically separated. Because of the physical dissociation of transmitters, the interference will be stronger depending on how close are the frequencies used. Therefore, one

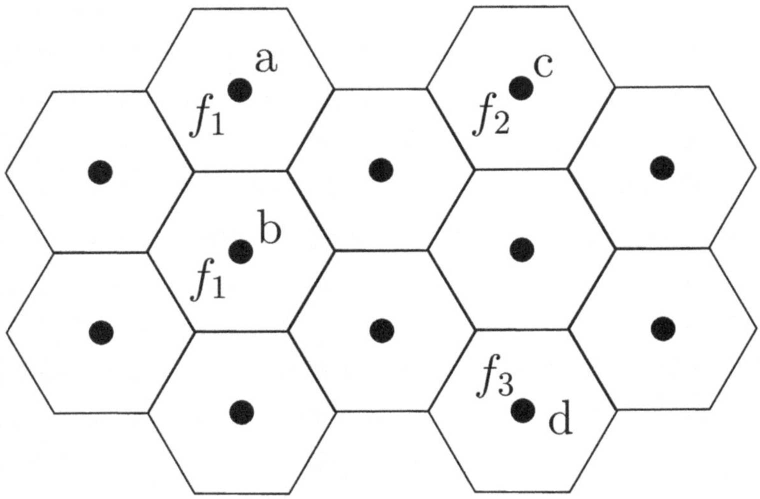

Fig. 6.4 Co-channel and adjacent-channel interference. Transmitters a and b use channel f_1. Transmitters c and d use, respectively, channels f_2 and f_3, which are adjacent in the frequency spectrum

can classify far-site interference based on the separation of frequencies used by transmitters.

When two transmitters in different locations use the same channel, this can result in a strong far-site interference, felt by most receivers of the signal. This kind of interference is called *co-channel* interference, as illustrated in Fig. 6.3.

An interference of lower intensity occurs when two transmitters in different locations use channels that are adjacent in the spectrum, that is, when the difference $|y(a) - y(b)|$, between the channels $y(a)$ and $y(b)$ assigned to transmitters a and b, is equal to 1. The amount of interference in each of these cases is empirically found for each pair of transmitters in the system and tabulated in the matrix $C_{i,j}$ of co-channel interferences. In the same way, one can create a matrix $A_{i,j}$ with observed values for adjacent-channel interference.

Figure 6.4 displays an example in which both co-channel and adjacent-channel interferences occur at the same time. Antennas a and b use the same channel, while antennas b and c use adjacent channels. The co-channel interference between b and c is given by the value $C_{a,b}$, while the adjacent-channel interference between b and c is given by the value $A_{b,c}$.

6.3.4 Determining Channel Separation

The amount of interference that is generated in one of the forms discussed in the previous section can have distinct impacts in the operation of mobile telephony

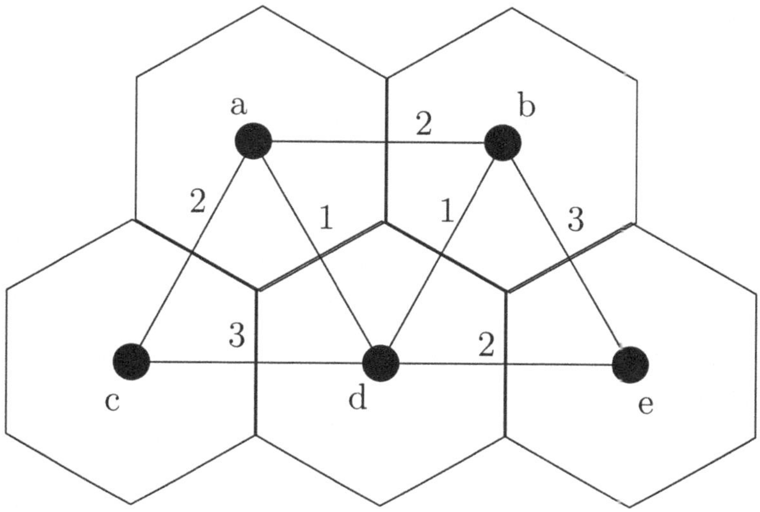

Fig. 6.5 Interference graph for a set of five cells

systems. Such interference can, in the best case, generate only a minimum amount of noise in the transmission. However, when interference crosses a certain threshold, the quality of service for the transmission can be severely compromised.

Diverse factors can contribute to the amount of interference observed in mobile devices. Some of the factors that serve to limit the acceptable amount of interference include:

- Sensibility of the devices
- Their capacity to filter undesirable frequencies
- The distance between transmitters.

Such limits are usually defined in an experimental way, such as to guarantee the normal operation of the mobile telephony system.

To avoid interferences surpassing the maximum acceptable limit, one can, for example, define a function $d(a, b)$ between antennas a and b, such that $d(a, b)$ represents the minimum distance between channels assigned to these antennas. For example, in Fig. 6.5 a set of cells is presented, with the respective separation limits between base stations being denoted by edges. When there is no edge between two base stations, this means that there are no distance restrictions between them. Therefore, transmitters located at b and c could receive any of the available channels. A graph in which the edges represent the minimum separation between frequency channels is called *interference graph*.

6.4 Types of Frequency Allocation

Frequency allocation is the assignment of frequency channels to each transmitter in a cellular telephony system. Depending on how it is performed, the frequency allocation can be either static or dynamic.

- **Static allocation** A static allocation of frequencies is one that is performed right after the mobile phone network is already designed or when its topology has been updated. Algorithms for static allocation can be, therefore, more elaborated and take relatively more time, since they need to be run only a few times during the lifetime of the mobile telephony system. Although such algorithms are allowed to execute for longer periods, due to the increasing size of instances and the complexity of the problem, there is still the need of using fast approximate algorithms, capable of producing high-quality solutions in an acceptable time.
- **Dynamic allocation** In contrast, dynamic allocation is performed as soon as new channels need to be added to a base station. This may happen, for example, when the number of users accessing a base station exceeds its service capacity. If more channels are available in other stations, one can try to reuse them to serve the extra demand. Algorithms for dynamic allocation of channels need to work in a distributed way and in real time. Moreover, since they need to consider several dynamic characteristics of the cellular technology in which they are used, such algorithms are much more tied to a particular standard or technology. A lot has been published in this area in the last years [2, 9]. In Katzela and Naghshineh [4], a detailed discussion is provided on the aspects of static and dynamic allocation.

Other than these two main types of algorithms for frequency allocation, work has also been done on other methods that use intermediary strategies. An example is the borrowing channel assignment strategy (BCA) [5]. In this method, cells have a fixed channel assignment. At the moment in which the number of requests becomes greater than the capacity of the cell, additional channels can be borrowed from other cells. In Sandalidis et al. [8], BCA is modeled as an optimization problem, and three heuristics are presented, based on genetic algorithms and neural networks.

Despite the convenience of dynamic channel allocation, one needs to note that most of the time it presupposes the use of a static allocation as its basis. Therefore, the quality of static allocation has great importance in defining the general results of a mobile telephony allocation strategy. In this book, we will discuss static allocations of channels exclusively. Unless otherwise indicated, one should understand this as the standard type of frequency allocation method used in our discussion.

Due to the minimum channel separation $d(a, b)$, required between each pair of transmitters a and b in a mobile phone network, one can classify assignments as valid or invalid. All frequency channel allocations that do not respect the minimal channel separation are considered to be invalid, since they violate required constraints. In formal terms, a frequency allocation is considered to be valid if, for each pair (a, b) of transmitters in the system, the difference between the frequencies assigned to a and b is less than $d(a, b)$.

6.5 Mobile Technologies

Throughout the last decades, several technologies have been used in mobile telephony. The main reason for this growth is the rapid advance observed in this area, with several companies competing for dominance in a market that still lacks standardization in some of the forefront areas. In this section we will discuss some of the key technologies that provide similar services in the sector of mobile telephony and what they have in common.

6.5.1 TDMA

When a frequency channel is used for voice transmission, it can be noticed that, during several time periods of pause, it is not being actively used to transfer data. For example, when one of the participants in the conversation is talking, the other participant is generally listening and its device is only receiving data. This means that, during most of the time of a normal call, the communication channel is being used in a unidirectional way.

Another fact about voice channels is that, even during a typical conversation, there are lots of small pauses during which no data is being transferred. Therefore, it is possible to make better use of the communication channel if we devise a way to use these periods, when no data is transferred, for other purposes. These natural features of voice channels are used by mobile telephony systems such as TDMA to optimize the transmission of data during a typical conversation.

The TDMA (time-division multiple access) standard uses a characteristic of the human ear perception as the basis for its operation. This characteristic concerns the fact that humans cannot hear changes in sound that occur too quickly. This indicates that, instead of transmitting each and every sound change, it is enough to send voice data only within small intervals, such as once each 10 ms.

Noticing these characteristics of voice transmissions, a set of strategies were developed by the creators of TDMA. These strategies have the goal of drastically reducing the amount of data that needs to be transferred during a phone call. When applied to mobile telephony, these methods allow that a frequency channel be shared by several users at the same time, without noticeable lack of quality in the resulting transmission.

At its core, TDMA is a technology that divides a frequency channel into several slots. The division allows channels to be shared by several transmissions in parallel. The division is made according to a predefined time slices, and existing calls are assigned specified periodic slots in these time slices. In Fig. 6.6 one can see an example of frequency sharing using time slots. Three users, represented as u_1, u_2, and u_3, send information in turns using slots s_1, s_2, and s_3, which divide the unique channel that is shared by these users.

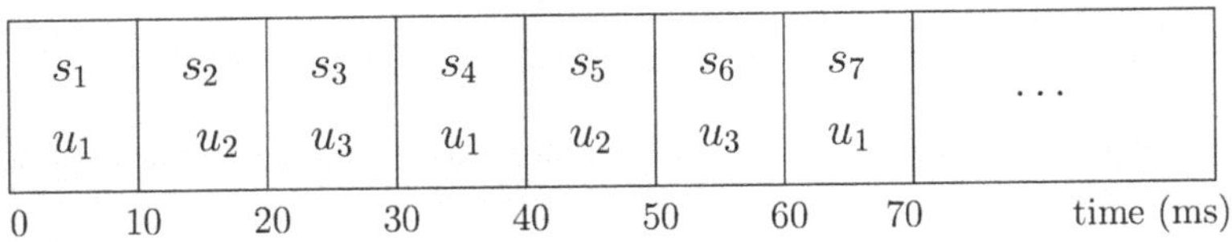

Fig. 6.6 Division of a frequency channel into time slots

While TDMA is an older generation of cellular technology, it is still used in combination with other techniques for frequency spectrum management. As such, TDMA is currently the basis for several technologies that are employed on cellular networks. It may also be used in combination with CDMA, as discussed in Sect. 6.5.4.

6.5.2 AMPS and D-AMPS

AMPS (advanced mobile phone system) is a mobile communication standard developed for the needs of cellular networks. It is one of the oldest systems and was the first to be used in regions such as North America, Latin America, and the Middle East. AMPS appeared in the US market in 1983 and was quickly popularized. In its initial version, AMPS used analog technology which was adequate to the needs of its time. Gradually, however, digital techniques were added to AMPS so that it could compete with other cellular network techniques. The digital version of AMPS is also known as D-AMPS.

Being a pioneering technology also means that AMPS had some disadvantages compared to more recent cellular standards. One of the disadvantages of the original AMPS is related to the smaller number of users that could be served by a single cell. Also, there were problems with limitations on the use of the frequency spectrum, in the safety of data transmission (e.g., data security and integrity could be compromised) and difficulties with handoff operations, which could cause intermittent service.

Despite this, one of the main advantages of AMPS for decades was its simple and easy implementation, in relation to more complex technologies. This made to become widely accepted, having at a time corresponded for a large percentage of mobile cellular phone systems in use throughout the world.

As a way to rehabilitate AMPS, a new version of this standard was developed incorporating new features. This new version, called D-AMPS, supports the use of more advanced techniques such as FDMA (frequency-division multiple access) and TDMA (time-division multiple access). D-AMPS was targeted specifically at systems that wished to make an easy transition between analog and digital transmission methods.

6.5.3 GSM

The GSM (Global System for Mobile Communications) was created with the specific purpose of becoming the European standard for mobile telephony [3]. Among the main characteristics of the original GSM standard are:

- Good quality of voice transmission
- Low cost for the implementation and operation of the service
- Direct support for roaming on areas not covered by the original carrier
- Support for handheld devices
- More efficient use of the frequency spectrum
- Compatibility with the existing ISDN (Integrated Services Digital Network) standard

Initial research for the creation of the GSM started in 1982. Proposed by a commission of the Conference of European Posts and Telegraphs, the GSM standard was finally adopted by the European Telecommunications Standards Institute in 1989. The basic method used by the GSM for the purpose of signal codification is TDMA, while it can also use CDMA for this role. The GSM is a standard firmly established in Europe, which is also being used in several other regions of the world.

6.5.4 CDMA

CDMA (*code-division multiple access*) is a technique for multiplexing the access to a frequency channel. As such, it has been the subject of research for several decades. The first country to make investments in CDMA-based telecommunications was the USSR [1], where cellular networks were developed for military use.

The modern version of CDMA as a cellular telephony technology was developed in the USA, where the system has found its largest number of users. The current version of CDMA, developed initially by the Qualcomm Incorporated, offered many advantages over other systems used at the time. Its main feature is the complete use of the frequency band available for transmission, without the need of previous division into channels. This allows for multiplexing and the use of codes for the simultaneous transmission of data, making the system more robust and increasing the safety and secrecy even without the use of cryptography. Other advantages can be pointed out in the use of CDMA, including better control of transmission power, allowing a reduction in interference and decreasing the noise by carefully exploiting silences during a conversation, resulting in a bigger number of available channels.

Initially, CDMA was a complex and expensive system to implement, in practice restricting its use. However, due to its advantages, implementation costs have decreased as the technology matured and was adopted in other markets around the world.

6.6 Optimization Problems in Mobile Telephony

As the area of mobile telephony grows, the challenges of designing and operating these large networks also become more complex. As a result, many of the difficulties associated to mobile telephony have also become new challenges for operations research. Most of these challenges are related to discrete, combinatorial problems, which are subject to the computational issue of combinatorial explosion. For many of these tasks, the general structure corresponds to NP-hard problems for which no efficient algorithm is known in practice. Sometimes, it is difficult even to provide approximate solutions for such problems. Therefore, new research is necessary to find methods capable of handling the large instances found in practice.

In the area of mobile telephony, NP-hard problems can be identified in all phases, including design, implementation, and management. Many of such NP-hard problems are of fundamental importance for the efficient implementation of mobile networks. In the following sections, we present some examples of commonly occurring mobile telephony problems. They include the optimal location of base stations, power management, and mobile device handoff between mobile cells [6].

6.6.1 Base Station Location Problems

The problem of locating base stations consists in determining where these facilities can be built, in such a way that a given objective function is optimized. Common optimization objectives include:

- The cost of real estate used
- The amount of power needed for base station operation
- The number of users served by the set of base stations
- The maintenance cost of selected base stations

A number of restrictions must also be considered, such as the maximum and minimum distance between base stations and the minimum power level that is acceptable for the given specifications.

It is necessary to observe that the optimal location of base stations also influences a number of related decisions and activities. For example, the location of base stations may affect signal levels on different parts of the covered area due to geographical factors. This kind of decision needs to take into consideration different aspects of the terrain, such as existing obstacles, buildings, and surrounding landscape.

6.6.2 Power Management Problems

Power management is another problem that is an integral part of the planning and implementation of any mobile telephony system. Given a set of transmitters, one needs to determine the optimal power level to be used, with the specific goal of reducing interferences and providing a satisfactory level of service.

Finding the ideal power levels have important effects on the operation of the cellular network. If the power levels used are too low, mobile devices may suffer from recurrent signal loss. On the other hand, if power levels are too high, then transmitters will be subject to undesirable interference from neighboring antennas. To achieve the optimal value, several parameters need to be adjusted, including the distances between transmitters as well as the specific frequencies used.

6.6.3 Mobile Cell Handoff Problems

Another difficult problem in the area of mobile telephony is related to possible changes in the base station serving devices, while they move between cells—an event that is known as handoff. In a mobile telephony system, a cell corresponds to the region served by a set of antennas.

Consider the situation illustrated in Fig. 6.7, where a mobile device moves from cell a to cell b. When this happens, a number of measures need to be taken by the cell management system to guarantee connectivity to the mobile user. It may be necessary, for example, to allocate another frequency from the pool of available frequencies on the receiving cell. It is also necessary to indicate to the system that the traffic of data will flow to the antenna of base station b, instead of the previous antenna.

One can view the handoff as an optimization problem, where the objectives include minimizing the total time for reestablishment of communication. Another goal might be that of minimizing the interference caused by the allocation of a new frequency. Making the cellular handoff to happen in a way that is transparent for

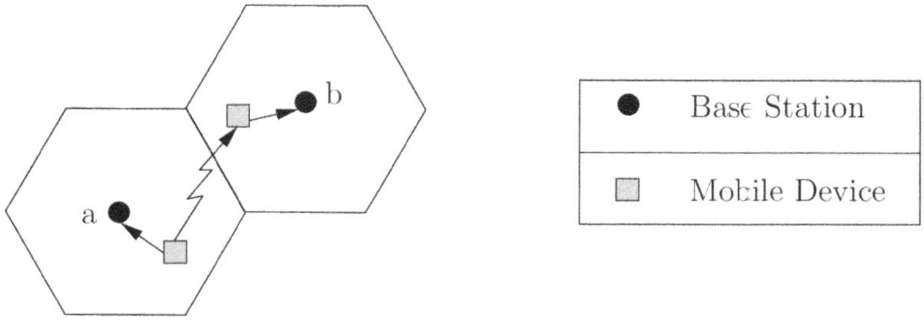

Fig. 6.7 A mobile device moves from cell a to cell b, resulting in a handoff

users, without loss of information, and by causing only a smooth transition for the whole system is a big challenge for designers of cellular systems.

6.7 Conclusion

In this chapter, an overview of routing problems in mobile telephony systems is presented, including the technology and basic terminology used in this area. To have a better grasp of the optimization problems that will be studied in the next chapters, it is important to become familiarized with the basic concepts of cellular systems.

The basic technology used in cellular system was introduced, along with some of its historical background. We have also presented some of the most commonly found optimization problems in this area. In a future article we present general optimization concepts and strategies that are used in the solution of frequency assignment, routing, and related optimization problems.

References

1. D.V. Ageev, Bases of the theory of linear selection. code demultiplexing, in *Proceedings of the Leningrad Experimental Institute of Communication* (1935), pp. 3–35
2. G. Cao, M. Singhal, *Efficient Distributed Channel Allocation for Mobile Cellular Networks* (1997). Department of Computer and Information Science, The Ohio State University
3. F.J. Fahham, Cellular communication networks—GSM, in *Surveys and Presentations in Information Systems Engineering* (Department of Electrical and Electronic Engineering, Imperial College of Science Technology and Medicine, UK, 1996)
4. I. Katzela, M. Naghshineh, Channel assignment schemes for cellular mobile telecommunication systems. IEEE Pers. Commun. Mag. **3**(3), 11–31 (1996)
5. C.Y. Lee, *Mobile Cellular Telecommunications Systems* (McGraw-Hill, New York, 1989)
6. G.R. Mateus, A.F. Loureiro, *Introdução à Computação Móvel*, in 11^a *Escola de Computação, Rio de Janeiro* (1998)
7. R. Periannan, Existing technology of cellular networks—mobile phones, in *Surveys and Presentations in Information Systems Engineering, Department of Electrical and Electronic Engineering, Imperial College of Science Technology and Medicine, UK* (1996)
8. H.G. Sandalidis, P.P. Stavroulakis, J. Rodriguez-Tellez, Borrowing channel assignment strategies based on heuristic techniques for cellular systems. IEEE Trans. Neural Netw. **10**(1), 176–181 (1999)
9. S. Singh, D. Bertsekas, Reinforcement learning for dynamic channel allocation in cellular telephone systems, in *NIPS96* (1996)

Chapter 7
AI Techniques for Combinatorial Optimization

Carlos A. S. Oliveira

Abstract Optimization problems are prevalent across various industrial sectors such as agriculture, transportation, and administration, where the goal is to efficiently allocate resources to achieve specific objectives. These problems often involve minimizing or maximizing a function subject to constraints, and they can be mathematically formalized as a set of equalities and inequalities. Depending on the nature of the objective function and constraints, optimization problems can be classified into different categories, including linear programming and combinatorial optimization. This paper introduces the fundamental concepts of optimization problems, provides examples of real-world applications, and delves into specific problem types such as the Traveling Salesman Problem (TSP) and the Knapsack Problem. The TSP, a well-known combinatorial optimization problem, seeks the shortest possible route that visits a set of cities exactly once and returns to the origin, with applications ranging from circuit design to logistics. The Knapsack Problem, another classic example, involves selecting items with maximum value within a given capacity, with practical applications in resource allocation and investment decisions. Despite their simple formulations, both problems pose significant computational challenges, particularly when finding optimal solutions for large instances.

Keywords Optimization · Metatheuristics · TSP · Tabu search · Genetic algorithms

7.1 Introduction

For a large number of industrial problems, a set of resources need to be used in such a way as to achieve a preestablished goal. This happens in important sectors of

C. A. S. Oliveira (✉)
Principal Inventive Scientist, Network Chief Technology Organization, AT&T Labs Inc., Bellevue, WA, USA
e-mail: carlos.oliveira@att.com

C. A. S. Oliveira, M. P. Pardalos (eds.), *Handbook of Artificial Intelligence and Data Sciences for Routing Problems*, Springer Optimization and Its Applications 219, https://doi.org/10.1007/978-3-031-78262-6_7

the economy such as agriculture, transportation, and administration, just for a few examples.

For instance, one may want to manufacture a product while at the same time hoping to use the minimum amount of raw material. This would naturally lead to a combinatorial problem where the optimal combination of raw materials is desired. For an concrete example occurring in the healthcare industry, one may want to identify the best location for hospitals and service ambulances, in such a way that the local population can receive assistance in the minimum amount of time after an occurrence.

What these problems have in common in the presence of restrictions, imposed by economic or social constraint, and solved in order to achieve a goal that can be either the minimization or maximization of a function. Problems such as the ones described above are classified as *optimization problems*.

A generic optimization problem can be established as a set of equalities and inequalities of the following form:

optimize f(x)
subject to: $g_i(x) \geq 0, \quad i = 1, \ldots, m$
$h_j(x) = 0, \quad j = 1, \ldots, n$

In this set of equations, $f(x)$ is called the *objective function* and represents the value that must be optimized (maximized or minimized). Functions $g_i(x)$, $i = 1, \ldots, m$ e $h_j(x)$, $j = 1, \ldots, n$ are referred to as *problem restrictions* and specify particular properties that a feasible solution must possess in order to be acceptable. For example, the solution to an oil production model has to satisfy constraints related to the quantity and type of raw materials used in the process.

A *feasible solution* for an optimization problem is a set of values $a_1, a_2, a_3, \ldots$ for the variables $x_1, x_2, x_3, \ldots$ appearing on the preceding equations, such that $g_i(x)$, $i = 1, \ldots, m$, and $h_j(x)$, $j = 1, \ldots, n$ are satisfied.

An instance of an optimization problem Π is a pair (C, f), where C is the set of all feasible solutions for Π and $f : C \to \mathbb{R}$ is the objective function. Then, an optimization problem can be alternatively defined as finding an element $x \in C$ such that $f(x)$ is the largest (or smallest) possible.

7.2 Classification of Optimization Problems

Depending on the format of functions f, g_i and h_j, one can classify optimization problems into some general families. The most interesting cases happen when the objective function and its restrictions are

1. *Convex*, that is, they have only one local (and therefore global) minimum (or maximum);
2. *Linear*, that is, they are formed only by the sum of the scalar product of variables.

Variables appearing on functions f, g_i e h_j are also called *decision variables*, because they affect and define the value of the objective function.

Optimization problems that are composed only of linear functions are also known as *linear programming* problems. They can be represented in the following way:

$$\textbf{minimize} \sum_{i=1}^{n} c_i x_i$$

$$\textbf{subject to:} \sum_{i=1}^{n} A_{j,i} x_i = b_j, \text{ for } j = 1, \ldots, m$$

$$x_i \geq 0, \text{ for } i = 1, \ldots, n$$

where Aj, i represents an element of the weight matrix, x_i is a decision variable, and b_i and c_i are elements of vectors with dimension m and n, respectively.

An important class of optimization problem occurs when decision variables are restricted to integer values. A problem of this type has an integer number (finite of infinite) of objects. In this case, the problem is then known as combinatorial optimization (or discrete optimization). These types of problems have been widely studied due to the number of applications in real-life problems.

Combinatorial optimization problems vary in difficulty and can be classified according to the number of computer instructions necessary to solve them. Some of these problems, such as the network flow problem, already have known solution procedures that can be executed in time that is bounded by a polynomial on the input size. In other words, the execution time can be expressed as a polynomial, where variables represent the number of bits necessary for coding all data for the problem [10]. However, for a large class of combinatorial optimization problems, no algorithm with such performance has been found to this date. This class of problems is known as NP-Hard, and finding optimal solutions for their members is one of the biggest challenges for researchers in theoretical computer science (see [4]).

As an illustration of the different types of combinatorial optimization problems, I present next a few classical examples. They are the Travelling Salesman Problem (TSP) on Sect. 7.3), the knapsack problem on Sect. 7.4, and the problems of maximum independent set and maximum clique on Sect. 7.5.

7.3 The Traveling Salesman Problem

The Traveling Salesman Problem (TSP) is one of the best known problems in combinatorial optimization. Informally, it can be described as finding the fastest way to visit a set of points in a plane, such that we return to the first point and all points are visited only once.

If we consider the points as being city locations, the solution for this problem may represent the best path for a sales person who want to visit all of the n cities in the input set. This is the original context in which the problem was studied. In

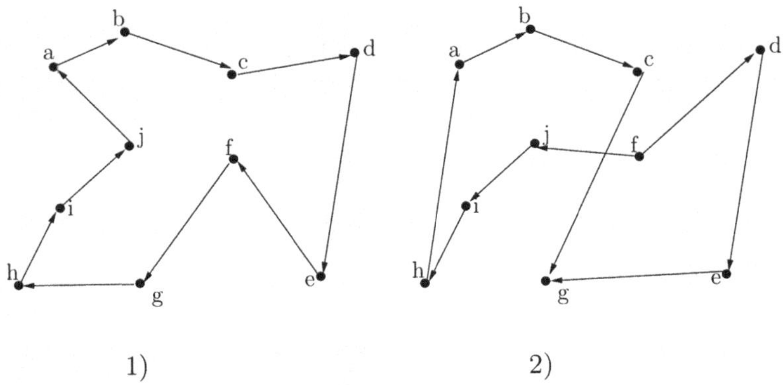

Fig. 7.1 Instance of the Traveling Salesman Problem, with ten points. Two feasible solutions are presented as (1) and (2). The total travelled distance on the first path is 182 units, while in the second path the distance is 195 units

Fig. 7.1, we see two possible paths. It can be seen that the first path is smaller than the second path.

A simplified mathematical explanation of the problem can be provided in the following way: Give a set P of points in the plan; find a sequence $\{p_1, p_2, \ldots, p_n\} \subset P$ of points to be visited only once, returning to the original point p_1, while traveling the minimum distance on the plane.

A more formal definition of the TSP can be found, for example, on [3]. Here, we assume that distances are provided between each pair or points i and j, represented as $c_{i,j} \in \mathbb{R}$. In order to model the problem, decision variables $x_{i,j}$ are used. These variables indicate if point j is visited immediately after i ($x_{i,j} = 1$) in the visitation sequence. We then have the following formulation.

$$\textbf{minimize} \sum_{i=1}^{n} \sum_{j=1}^{n} c_{i,j} x_{i,j} \tag{7.1}$$

$$\textbf{subjectto} \sum_{j=1}^{n} x_{i,j} = 1, \text{ for } i = 1, \ldots, m \tag{7.2}$$

$$\sum_{i=1}^{n} x_{i,j} = 1, \text{ for } j = 1, \ldots, m \tag{7.3}$$

$$\sum_{i \in S} \sum_{j \in S} x_{i,j} \leq |S| - 1, \text{ for all } S \subset V \tag{7.4}$$

$$x_{i,j} \in \{0, 1\} \tag{7.5}$$

where $n = |P|$ is the number of points and S is a proper subset of $P = \{1, \ldots, n\}$. Function 7.1 establishes the objective, which in this case is minimizing the total distance. Equations 7.2 and 7.3 guarantee that each point has exactly one connection arriving to or leaving from it. Finally, Eq. 7.4 makes sure that the solution has no disconnected cycles (i.e., all paths are connected into a single cycle).

The TSP has a large number of applications. Among them, we can list construction of LSI circuits, job scheduling [7], and vehicle routing [2].

Despite the fact that the TSP seems to be a simple problem at first sight, no efficient algorithm was ever found to solve it. The simplest algorithms for the problem would verify each of the possible solutions and return the shortest one. However, starting from any point, there are $\frac{(n-1)!}{2}$ possible paths. Each new point represents a factorial increase in this number. For example, even a problem with just 10 points has more than 180,000 options that need to be explored, making this an impractical algorithm.

7.4 Knapsack Problem

In the knapsack problem, one is given a volume of c units, corresponding to the knapsack capacity. Using the available volume, one can hold any of the elements of the n-object set $X = \{x_1, x_2, \ldots, x_n\}$. Each object has volume t_i, for $i = 1, \ldots, n$, and value p_i, $i = 1, \ldots, n$. The objective of the problem is to find a configuration of objects with total volume of value at most c, and that maximizes the value hold of the knapsack. A mathematical formulation can be given as follows:

$$\textbf{maximize} \sum_{i=1}^{n} p_i x_i \tag{7.6}$$

$$\textbf{subjectto} \sum_{i=1}^{n} t_i x_i \leq c, \text{ for } j = 1, \ldots, m \tag{7.7}$$

$$x_i \in \{0, 1\}, \text{ for } i = 1, \ldots, n \tag{7.8}$$

where each x_i is a decision variable with value 1 if object i is contained in the knapsack, and value 0 otherwise.

The practical applications in which the knapsack problem occurs can be mainly found in industrial processes, on sectors as diverse as manufacturing and in decision-making. For example, on management, one may want to model the situation in which there is a fixed amount of money to implement improvements in a company. Each improvement has a defined cost. The goal is to maximize the financial return associated with the invested capital. In this case, the investments will be modeled by a set $\{x_1, \ldots, x_n\}$, where each investment has a cost and a given return. At the end of the optimization process, the total cost of the investments cannot exceed the available capital, which corresponds to the value c.

7.5 Maximum Independent Set

The maximum independent set problem (MISP) is a typical problem occurring in graph theory (as will be describe in more detail in Sect. 8.4). A *graph* $G = (V, E)$ is composed of a set $V = \{v_1, \ldots, v_n\}$ of *vertices* and a set $E = \{\{v_{i_1}, v_{j_1}\}, \ldots, \{v_{i_m}, v_{j_m}\}\}$ of pairs of vertices in V. The unordered pairs of vertices in $\{v_i, v_j\} \in E$ are also called edges. Given a graph $G(V, E)$, the goal is to find a maximum subset of vertices $V' \subset V$ such that for each pair of vertices $(x, y) \in V'$, there is no edge between x and y, i.e., $(x, y) \notin E$. Figure 7.2 depicts an instance of the problem. The vertices in black represent an independent set.

The MISP is used to model situations in which one needs to find the maximum set of elements that contains no particular relationship between each other. For example, in a group of people, finding the maximum subset of those people that don't know each other is an example of MISP.

7.6 Maximum Clique Problem

The maximum clique problem (MCP) is another graph-theory problem that is related to the MISP [4]. MCP, which is also an NP-hard problem, has the following objective. Given a graph $G(V, E)$, find the largest subsect $V' \in V$ that is complete.

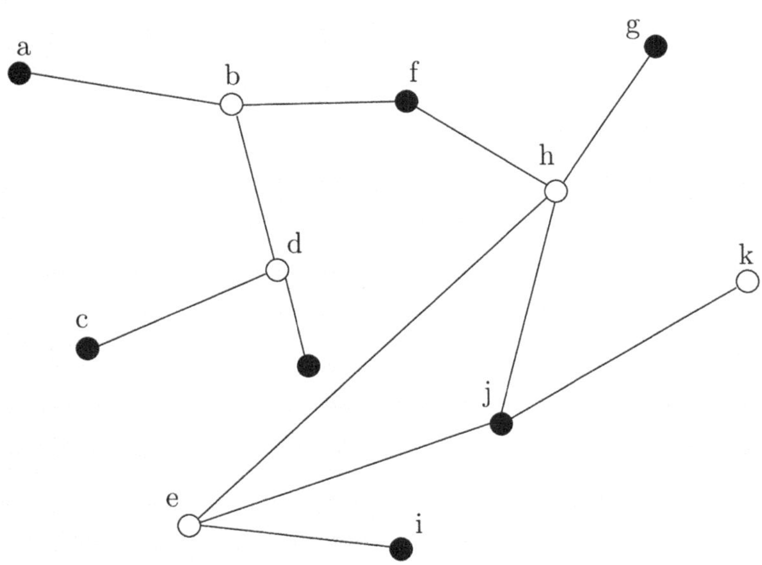

Fig. 7.2 Instance of the maximum independent set problem

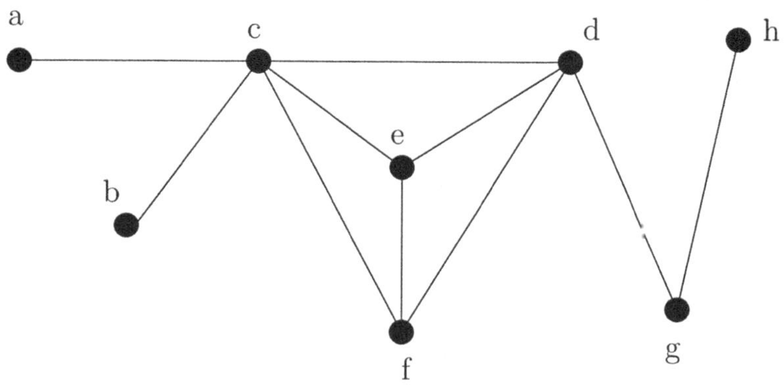

Fig. 7.3 Instance of maximum clique. Here, the maximum clique is the subgraph induced by c, d, e and f

In other words, for each pair of vertices $(x, y) \in V'$, there is an edge $(x, y) \in E$ in G. Figure 7.3 presents an example of clique, formed by vertices c, d, e and f.

Practical applications of the MCP can be found in the most different areas. One of the many situation would be, in telephonic carrier, finding the largest group of users that called all other group members in the last month. In this case, each user is represented by a vertex in the graph. Edges exist between pairs of users whenever a call has been made between them during the specified period.

The relationship between the maximum clique problem and the maximum independent set is based on their complimentary characteristics. Give a graph $G = (V, E)$, the complement of G is denoted by $\overline{G}(V, E')$. The complement graph is defined as having the same set of vertices as the original graph, but with edges $(x, y) \in E'$ if and only if $(x, y) \notin E$. Using this definition, one can easily observe that by finding the maximum clique of G, one is also determining the maximum dependent set of $\overline{G}$. This implies that both problems have the same time complexity [4].

7.7 AI Problems and Optimization in Transportation

AI is reshaping transportation networks, making them more efficient, safer, and environmentally friendly. However, implementing these solutions requires overcoming data, safety, and ethical challenges. Transportation optimization is a multifaceted problem, and AI technologies are providing innovative ways to tackle its complexities. The integration of AI into transportation networks is a dynamic field, with ongoing research and development pushing the boundaries of what is possible. As AI systems continue to mature and adapt to new challenges, we can expect transportation networks to become more responsive, sustainable, and user-friendly.

In an era of rapidly growing urbanization and increasing demands on transportation infrastructure, optimizing transportation networks has become a critical task for governments, businesses, and individuals alike. Harnessing the power of Artificial Intelligence (AI) has emerged as a game-changer in addressing the complex challenges associated with transportation network optimization.

In this article, we explore the vital role of AI in transforming transportation networks, delving into the challenges faced and the innovative solutions that AI brings to the table. Transportation networks, including road systems, public transit, and logistics operations, have to deal with a myriad of challenges. Congestion, pollution, inefficient routes, and safety concerns are just a few of the pressing issues that need to be addressed. Traditional optimization methods, while useful, often struggle to adapt to the dynamic and complex nature of these problems. AI-based methods may benefit from the capacity of AI systems to embrace large-scale data in order to improve the quality of results for this class of optimization challenges.

7.8 AI Techniques for Transportation

AI's ability to process vast amounts of data, learn from it, and make real-time decisions has opened up new horizons in transportation optimization. Machine learning algorithms, deep reinforcement learning, and other AI techniques offer the potential to optimize routing, improve traffic flow, and enhance the overall transportation experience.

While AI presents an exciting opportunity, there are several challenges to overcome in implementing AI for transportation optimization. These include issues such as data quality and availability, safety and security of resulting transportation decisions, AI collaboration issues, as well as regulatory and ethical concerns.

From the point of view of data quality and availability, we notice that accurate and timely data is essential for AI algorithms to function effectively. In many cases, data is incomplete, outdated, or not readily available.

Another important consideration is safety and security of resulting transportation. Implementing AI in transportation networks requires robust cybersecurity measures to protect against potential threats and ensure passenger and cargo safety.

Developing interfaces and protocols that allow AI systems to collaborate effectively with human operators is also essential, particularly in self-driving vehicles and autonomous transportation.

Finally, one needs to face with regulatory and ethical issues of using AI. The development and deployment of AI in transportation networks raise a host of ethical and regulatory concerns, such as privacy, accountability, and transparency. Here is a small number of innovative solutions for transportation that can be achieved with the help of AI algorithms:

- Real-Time Traffic Management: AI systems can analyze real-time traffic data from various sources, including GPS devices and traffic cameras, to optimize traffic signal timings, reduce congestion, and improve traffic flow. Adaptive traffic management systems have been successfully deployed in several cities, leading to noticeable improvements.

- Predictive Maintenance: In logistics and public transit, AI can predict maintenance needs for vehicles and infrastructure, reducing downtime and ensuring safer operations.
- Route Optimization: AI-driven route optimization algorithms help delivery companies plan the most efficient routes, minimizing fuel consumption and delivery times.
- Demand-Responsive Transit: AI-powered systems are making on-demand and flexible public transit services a reality, adapting routes and schedules in response to passenger needs and traffic conditions. Autonomous Vehicles: Self-driving cars and autonomous trucks, powered by AI, are poised to revolutionize transportation by reducing accidents and optimizing traffic flow.
- Multimodal Transportation: AI can facilitate seamless transitions between different modes of transportation, from buses to bikes to trains, offering a more efficient and ecofriendly transportation experience.

Artificial Intelligence (AI) plays a pivotal role in transforming and improving various aspects of transportation. In the next sections, we explore several examples of how AI and optimization are making transportation more efficient, safer, and sustainable. In the next sections, we consider these opportunity areas and explore the possible avenues for AI-based systems to improve existing methods.

7.9 Traffic Management

Traffic management allows the large flow of vehicles in a modern urban area [8]. It consists of several tasks that can be formulated as optimization problems. AI techniques can enhance the deployment of such traffic management systems.

A particular area for such efforts is real-time traffic analysis: AI algorithms can be used to analyze data from various data sources, including traffic cameras, GPS devices, and sensors. These data inputs can be used to monitor traffic conditions in real time. The analysis of this signal can be properly done by a well-trained AI system [9]. This information is then used to optimize traffic signal timings and reduce overall congestion.

Another field that benefits from the interplay of optimization and AI is predictive traffic modeling [12]. With these AI models, the goal is to predict traffic patterns and bottlenecks before they happen, and helping authorities and drivers to plan routes more efficiently.

Finally, we can cite dynamic pricing as an important challenge that can be tackled with AI systems. Ride-sharing platforms already use AI strategies to adjust prices, using signals based on demand, time, location, and other features [11]. This, in turn, works to encourage users to travel during off-peak hours or carpool.

7.10 Autonomous Vehicles

Autonomous vehicles have emerged as an important frontier of transportation systems [6]. Using sophisticated autonomous driving methods based on IA, these systems have the potential to transform the current urban landscape [5].

In the area of autonomous vehicles, one important goal is the deployment of self-driving cars and trucks. These AI-driven autonomous vehicles use sensors and machine learning to navigate and make real-time decisions, reducing accidents and improving road safety.

Another important area is that of enhanced ride-hailing services. Ride-hailing companies have invested large resources in AI research toward developing autonomous ride-sharing services. These services aim at making transportation more accessible and cost-effective for large segments of the population.

For example, Bischoff et al. [1] discuss the optimization of service areas for pooled ride-hailing operators, focusing on the case of Berlin, Germany. The authors propose a simulation-based methodology using an agent-based simulation (MATSim) to optimize ride-hailing service areas. Three optimization criteria are considered: average vehicle occupancy, revenues collected per area, and a combination of both. The study aims to determine economically justified areas for offering pooled rides.

Dynamic ride-hailing services, especially with passenger pooling, have gained popularity worldwide, disrupting traditional taxi markets. Pooled services generally operate in smaller, densely populated areas compared to non-pooled services. The authors [1] use MATSim, a multi-agent transport simulation, to model the taxi demand in Berlin and evaluate different optimization criteria.

Results show that, for the given parameters, a service area focusing on an extended central area and some surrounding zones may maximize profits for operators. The paper emphasizes the importance of considering both vehicle occupancy and revenues in optimizing service areas. The proposed methodology provides a framework for ride-hailing operators to strategically plan their service coverage based on real-world demand data.

7.11 Public Transit Optimization

Public transit systems benefit greatly from improved solutions to optimization problems using AI. A set of problems in public transportation require large amounts of resources to be used in concert to improve quality and reduce public costs.

One particular area is demand-responsive transit. In such an area, AI helps in optimizing public transit routes based on passenger demand and real-time traffic conditions. The goal is to offer more convenient and efficient services for the population that depends on these transportation services.

Another initiative that benefits from the interplay of AI and optimization is the design and creation of smart bus stops. With AI-powered bus stops, the goal is to provide real-time information about bus arrivals and service updates, making public transit more user-friendly.

7.12 Routing and Navigation

Routing optimization is a central task in any transportation system, especially for the development of supply chains and related economic activities. The study of routing has morphed into a dynamic field that encompass several traffic subsystems.

First, we have the important task of efficiently solving optimization models for routing problems. With AI algorithms, it has become possible to ship real-time navigation features as part of navigation apps, which can be used by travelers to determine the most efficient routes. These algorithms are designed in such a way as to take into account traffic conditions, road closures, and other real-time data.

Routing optimization also plays an important role in multimodal transportation. The use of AI systems in this area enables the seamless integration and transition between different modes of transportation. The typical example of such as system combines modes of transportation for end users, such as buses, bikes, and trains, for the completion of a single journey.

7.13 Fleet Management

Predictive maintenance is a vast area where optimization tools are used to determine the best decisions in order to maintain an existing feet of vehicles. Typical decisions contemplated in this situation is when to perform maintenance, which sequence of vehicles should be chosen for maintenance, and what kind of maintenance needs to be performed.

AI algorithms can also be used to predict maintenance needs for a fleet of vehicles while at the same time minimizing downtime and reducing repair costs.

Finally, an important decision for fleet managers is the choice that maximizes fuel efficiency. AI techniques can be used to help optimize fuel consumption in a fleet. This can be done in several ways, for example, by monitoring driving behavior, suggesting fuel-efficient routes, and reducing unnecessary idling, among other fuel saving measures.

7.14 Safety and Accident Prevention

AI algorithms and optimization can be also used to improve daily activities and increase safety on the transportation network. This can be done with algorithms for safety and accident prevention.

One possible area for accident prevention is the implementation of advanced driver assistance systems (ADAS). Closely related to self-driving algorithms, these AI-powered ADAS features are added to vehicles in order to provide assistance with common tasks, such as lane-keeping, adaptive cruise control, and collision avoidance. These AI-based techniques can be combined to maximize the safety of passengers.

Another area where AI algorithms can be used is in pedestrian detection and collision avoidance. In this case, AI systems in vehicles can be used to detect pedestrians, cyclists, and possible obstacles on the road. This analytical ability can also be used to enable automatic braking and more advanced collision prevention mechanisms.

7.15 Ecofriendly Transportation

Ecological concerns are an important consideration for modern government and industry. This is also true for transportation problems, and AI can be used to solve many optimization goals linked to sustainable transportation.

One of these tasks is the optimization of electric vehicle (EV) charging locations. With the help of AI, the goal is to optimize the charging of electric vehicles. This can be done, for example, by predicting customer demand and managing charging station availability accordingly.

Route Planning for Fuel Efficiency: AI suggests routes that maximize fuel efficiency, reducing emissions in long-haul trucking and delivery services.

7.16 Traffic Enforcement and Management

Finally, traffic enforcement is another area that is impacted by AI and optimization. A typical example of problems found in this area includes Automated License Plate Recognition (ALPR). Such AI-powered ALPR systems can be used to automatically recognize license plates, allowing for efficient enforcement of traffic laws. Other examples of problems that can use such techniques include toll collection and the tracking of stolen vehicles.

Intelligent traffic management of smart cities is just another example of what can be done using AI techniques for optimization. In this case, AI is a core component of smart city projects, where traffic signals, lighting, and other infrastructure are optimized. Common goals for such optimization projects include the improvement of traffic flow, increased safety, and maximal energy efficiency.

Finally, one can also use AI methods to tackle public transport predictive maintenance. Smart cities projects employ such AI algorithms to predict maintenance needs for buses, trains, and trams, as shown in Sect. 7.13, ensuring reliable and punctual service.

References

1. J. Bischoff, I. Kaddoura, M. Maciejewski, K. Nagel, Simulation-based optimization of service areas for pooled ride-hailing operators. Procedia Comput. Sci **130**, 816–823 (2018). The 9th International Conference on Ambient Systems, Networks and Technologies (ANT 2018)/The 8th International Conference on Sustainable Energy Information Technology (SEIT-2018)/Affiliated Workshops
2. S. Christofides, Vehicle routing, in *Traveling Salesman Problem*, ed. by E.L. Lawler, J.K. Lenstra, A.H.G. Rinnooy Kan, D.B. Shmoys (Wiley, New York, 1985)
3. G.B. Dantzig, D.R. Fulkerson, S.M. Johnson, Solution of a large scale travelig salesman problem. Oper. Res. **2**, 393–410 (1954)
4. M.R. Garey, D.S. Johnson, *Computers and Intractability—a Guide to the Theory of NP-Completeness* (W. H. Freeman and Company, New York, 1979)
5. S. Gong, J. Shen, L. Du, Constrained optimization and distributed computation based car following control of a connected and autonomous vehicle platoon. Transp. Res. B Methodol. **94**, 314–334 (2016)
6. M. Hyland, H.S. Mahmassani, Dynamic autonomous vehicle fleet operations: Optimization-based strategies to assign avs to immediate traveler demand requests. Transp. Res. Part C Emerging Technol. **92**, 278–297 (2018)
7. G. Laporte, The traveling salesman problem: an overview of exact and approximate algorithms. Eur. J. Oper. Res. **59**(2), 231–247 (1992)
8. M. Lippi, M. Bertini, P. Frasconi, Short-term traffic flow forecasting: An experimental comparison of time-series analysis and supervised learning. IEEE Trans. Intell. Transp. Syst. **14**(2), 871–882 (2013)
9. Y. Lv, Y. Duan, W. Kang, Z. Li, F.-Y. Wang, Traffic flow prediction with big data: A deep learning approach. IEEE Trans. Intell. Transp. Syst. **16**(2), 865–873 (2015)
10. C.H. Papadimitriou, K. Steiglitz, *Combinatorial Optimization: Algorithms and Complexity* (Dover Publications, Inc., New York, 1982)
11. E.I. Vlahogianni, M.G. Karlaftis, J.C. Golias, Short-term traffic forecasting: Where we are and where we're going. Transp. Res. Part C Emerging Technol. **43**, 3–19 (2014). Special Issue on Short-term Traffic Flow Forecasting
12. H. Yuan, G. Li, A survey of traffic prediction: from spatio-temporal data to intelligent transportation. Data Sci. Eng. **6**, 63–85 (2021)

Chapter 8
Telecommunication Networks and Frequency Assignment Problems

Carlos A. S. Oliveira

Abstract The Frequency Assignment Problem (FAP) is a critical issue in telecommunications, involving the optimal allocation of frequencies to transmitters under various constraints. This chapter provides a comprehensive overview of the FAP, starting with a formal definition and highlighting its complexity, including its classification as an NP-hard problem. The relationship between FAP and graph coloring is explored, showcasing how graph theoretical approaches can be applied to model and solve the problem. Various optimization objectives, such as minimizing interference and spectrum usage, are discussed. The chapter also reviews related work, emphasizing the evolution of algorithms and methods, particularly in the context of cellular networks, where minimizing interference is paramount. Finally, alternative formulations and advanced heuristics for solving the FAP are examined, offering insights into both theoretical and practical aspects of this complex problem.

Keywords Telecommunications · Frequency assignment · Assignment problem

8.1 Introduction

The frequency assignment problem (FAP) is a practical problem occurring in the area of telecommunications. For this reason, the FAP has been subject of research under several aspects, with various emphases depending on the particular application that is being targeted. In this chapter, we present a general definition of the frequency assignment problem. We will also provide an overview covering some of the proposed solutions to the problem.

This chapter is divided as follows. In Sect. 8.2, we provide a formal definition of the frequency assignment problem along with related definitions. Section 8.3 is

---

C. A. S. Oliveira (✉)
Principal Inventive Scientist, Network Chief Technology Organization, AT&T Labs Inc., Bellevue, WA, USA
e-mail: carlos.oliveira@att.com

C. A. S. Oliveira, M. P. Pardalos (eds.), *Handbook of Artificial Intelligence and Data Sciences for Routing Problems*, Springer Optimization and Its Applications 219, https://doi.org/10.1007/978-3-031-78262-6_8

dedicated to the complexity issues involving the frequency assignment problem. An important point for the correct understanding of the FAP is its relationship with the graph coloring problem. This will be the subject of Sect. 8.4. In section, we give an introduction to literature results, covering several aspects that are of importance for the study of the frequency assignment problem. Finally, in Sect. 8.6, we discuss alternative formulations for the FAP that have been proposed in the literature.

8.2 Problem Definition

In general, the frequency assignment problem has been used to designate a set of related problems, occurring in the broad field of data transmission using radio frequencies. These problems have in common the fact that one needs to allocate frequencies to transmitters in an optimal way. However, the restrictions imposed on the problem and the objective function can vary considerably. These variations depend on the architecture of the system and the types of interference to which the system is subject.

A classification for these problems was proposed by Hale [18]. Hale proposed a formalization of the frequency assignment problem using a graph theoretical model. Such formalization can be viewed as an extended version of the graph coloring problem. Graph coloring is classic problem in combinatorial optimization and extensively studied due to the fact that it is NP-hard. The graph coloring problem can also be used to model several other applications (see, e.g., [29]).

The frequency assignment problem can target different optimization objectives, as shown in [18]. One of these objectives is to minimize the number of channels used for frequency allocation. In other words, one wants to minimize $|\mathcal{X}|$, where

$$\mathcal{X} = \{x | y(c) = x \text{ for all } c \in C\}.$$

This objective is also known as minimizing the rank of the frequency assignment.

A second objective for this problem might be to minimize the difference $|y(b) - y(a)|$ between the numeric values for the largest assigned channel $(y(b))$ and the lowest assigned channel $(y(a))$. This difference is also known as the span and can be used to quantify the amount of spectrum that is used by the system.

Lastly, but with even greater interest for us, there is a third objective function that is used to minimize the total interference occurring in the system. Notice that the types of interference that are considered in a cellular telephony system are two: local interference, which happens between antennas that are located in the same base station, and adjacent channel interference, which happens between antennas using channels that are adjacent in the frequency spectrum.

8.2.1 Types of Interference

The focus of this work will be on the characteristics of the frequency assignment problem when applied to cellular mobile systems. The main features of the frequency assignment problem in this context are:

- The number of available frequencies for the use of cellular mobile telephony is predefined. Therefore, in general, it is not necessary to optimize the rank of the used channels or to minimize the span of the used frequencies in the spectrum. However, with the rapid increase in the number of users in mobile phone systems, the main objective has become to guarantee the existence of valid frequency assignments, with minimal interference. Therefore, the focus in the remaining chapters will be on minimizing the total interference resulting from channel allocations.
- The transmitters in a system are grouped in a few base stations, with known value assigned to local interference. This assumption makes it possible to treat frequency allocation inside a base station in a separate way.

Initially, we present a few definitions from graph theory that will be necessary for a mathematical formulation of the frequency assignment problem.

A graph $G = (V, E)$ is composed of a set $V = \{v_1, \ldots, v_n\}$ of *nodes* and a set $E = \{\{v_{i_1}, v_{j_1}\}, \ldots, \{v_{i_m}, v_{j_m}\}\}$ of pairs of nodes from V. The unordered pairs of vertices $\{v_i, v_j\} \in E$ are called *edges* or *arests*.

Let $G = (V, E)$ be a graph where V is the set of transmitters. Let $C \subset Z_+$ be the set of available channels. We will define two functions: $c^{co} : E \rightarrow [0, 1]$ indicates the level of co-channel interference and $c^{ad} : E \rightarrow [0, 1]$ indicates the amount of adjacent-channel interference between pairs of nodes. We also have $c^{co}(u, v) \geq c^{ad}(u, v), \forall u, v \in V$. A frequency allocation is a mapping y from V into C.

Given these definition, the frequency assignment problem can be defined as minimizing the function

$$f(x) = \sum_{y(v)=y(w)} c^{co}(v, w) + \sum_{|y(v)-y(w)|=1} c^{ad}(v, w) \tag{8.1}$$

such that

$$|y(v) - y(w)| \geq d(v, w) \tag{8.2}$$

for each pair of antennas $v, w \in V, v \neq w$.

Figure 8.1 shows an example of instance for the frequency assignment problem. The instance is composed on just four cells, named c_1, c_2, c_3, and c_4. Each cell has a base station, with a variable number of transmitters. The demand vector for base stations c_1, c_2, c_3, and c_4, respectively, is [4, 3, 3, 2]. In this picture, one can observe the interference graph formed by edges (c_1, c_2), (c_1, c_4), (c_2, c_3), (c_3, c_4), and (c_2, c_4). The minimal separation between channels for each pair of base stations

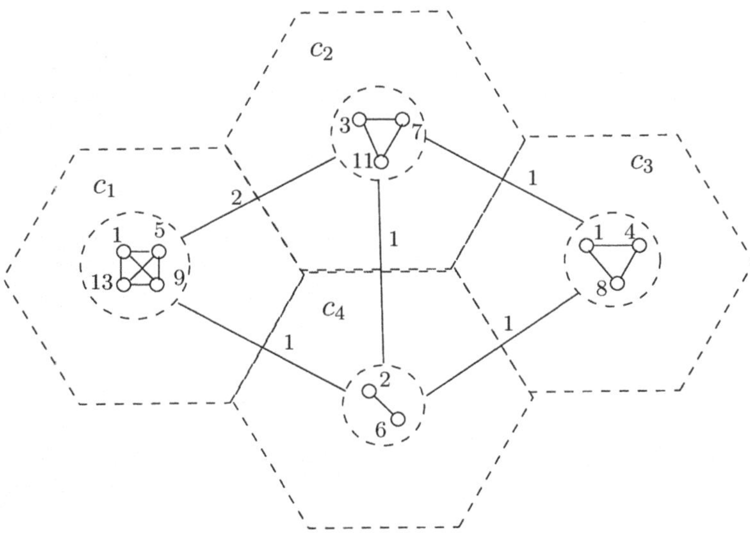

Fig. 8.1 Example of frequency assignment

is shown in the label of the corresponding edge. Finally, in the internal area of the base stations, one can see the number of the channel assigned to the respective transmitters. The frequency assignment presented here uses 11 distinct channels but occupies a range of 13 channels.

8.3 Problem Complexity

The first important results in terms of computational complexity for the frequency assignment are to determine the computational complexity class it is part of. This is stated in the following theorem.

Theorem 8.1 ([18]) *The frequency assignment problem is NP-hard.*

In order to show that the frequency assignment problem is NP-hard, it will be shown that the related decision problem (DFAP) is in the NP-complete class. We assume that all interference associated with the problem can assume only rational values. The DPAF, the decision problem associated with the FAP, is defined as follows. Given as input a minimal distance function $d : E \to N$, and the functions that represent co-channel and adjacent channel interference (c^{co} e c^{ad}), DPAF consists of answering yes or no to the following question: Is there a frequency assignment such that the resulting interference is less than or equal to an input value $x \in \mathbb{Q}$?

With this problem, for each algorithm that calculates a candidate answer, it is easy to verify if the answer is correct or not. Just calculate the sum of generated interference and verify that this sum is less than x. This can be done by just checking each edge of the interference graph G, which can be done in time $O(|E|) \leq O(|V|^2)$.

The last step of the proof is to reduce a problem Π', known to be NP-hard, to the DPAF. This step has been done in [18], using the k-coloration problem (a problem known to be NP-complete [17]) as the starting point.

Once the decision problem associated with the frequency assignment problem is proved to be part of the NP-complete class, one can then say by definition that the FAP is NP-hard.

8.4 FAP and Node Coloring

One of the ways of formalizing the frequency assignment problem is to treat the transmitters as nodes in a graph with edges for each pair of antennas in which interference may occur. The use of graph theory then becomes important for the correct understanding of the properties of this problem. Also, by doing this, existing results for related problems can be used in order to find algorithms for the FAP while still using properties unique to the problem.

In this section, we show how the frequency assignment problem can be modeled using ideas from graph coloring. Moreover, we will present some of the most important algorithms for graph coloring, which form the basis of the methods used in the solution of the FAP. Therefore, it becomes necessary to have a good understanding of such algorithms in order to discuss the solution methodology for the frequency assignment problem.

8.4.1 Graph Theory Definitions

In this section, we define some important problems in graph theory that will be useful when working with the frequency assignment problem.

Proper Node Coloring Given a graph $G = (V, E)$ and a function $w : V^2 \to \{0, 1\}$ defined as $w(a, b) = 1$ if $(a, b) \in E$, a proper coloring of G is a mapping $\phi : V \to N$ such that $|\phi(a) - \phi(b)| \geq w(a, b)$.

List Coloring Let $G = (V, E)$ be a graph with a set of nodes $V = \{v_1, \dots, v_k\}$. We denote by L_i the set of admissible colors assigned to v_i. The union of sets L_i is defined as

$$L = \bigcup_{i \in V} L_i.$$

A mapping $\phi : V \to L$ is a list coloring, or an L-coloring, if ϕ is a coloring for G and $\phi(v_i) \in L_i$ for all $i \in \{1, \dots, k\}$. In a few published papers, the lists L_i refer to disallowed colors. In this case, one can consider $L_i^{'}$ as the complementary relation applied to set L_i, which allows one to use the same concepts.

T-Coloring A T-Coloring is a special kind of coloring in which one avoids colors that are in given set. Given a graph $G = (V, E)$ and a finite set $T \subset \mathbb{N}$, a T-coloring is an assignment $y : V \to \mathbb{N}$ such that for each edge $(i, j) \in E$, $|f(i) - f(j)| \notin T$. It is possible to prove that the frequency assignment problem can be reduced to a T-coloring problem, as shown in [18]. This result shows one of the reasons for interest on this special type of coloring problem.

8.4.2 Heuristics for Graph Coloring

Many heuristics have been developed for the solution of the coloring problem. They are important in this area of study, because they may serve as the starting point for more elaborated heuristics for the frequency assignment problem. In this section, we discuss some of the most well-known heuristics for this problem [25].

The simplest methods for graph coloring start by verifying the nodes of the graph in sequence and assigning the first available color to each of them. Such algorithms are collectively known as sequential assignments. However, there may be variations in the sequence in which nodes are considered, leading to different versions of the algorithm. The method known as *largest first*, employed for example in [37], first sorts the nodes in ascending order according to their node degrees.

The method known as *smallest first*, used in [29], assumes that a similar ordering is applied, with the difference that the nodes are sorted as follows: First take the vertex with minimum degree, and put it at the end of the list. Then, from the remaining nodes, take the one with the smallest degree in the graph induced by the remaining nodes. Add the node selected in this way to the second-last position in the list. The algorithm will continue in this fashion until the list contains all nodes in the graph.

An improvement that can be obtained on the method discussed above is the performance of exchanges between nodes that have been previously colored, in a way that the new vertex can be colored with a previously used color. The methods that employ this type of exchange are called *largest first with exchanges* and *smallest first with exchanges*.

In [6], the DSATUR method was introduced, in which the nodes are selected according to their *saturation degree*. The saturation degree of a node is a quantity that changes during the algorithm and represents the number of different colors assigned to the neighbors of a given node. At each step, the node with highest saturation is selected, and a valid color is assigned to it. The idea behind such a procedure is that one should select first the nodes with fewer options with respect to the assignment of colors.

Algorithm 8 RLF procedure

$V \leftarrow$ the set of nodes that have been assigned a color.
$\overline{V} \leftarrow$ the set of nodes with no color assignment.
Select the node $v \in \overline{V}$ with highest degree in the graph induced by $\overline{V}$.
$U \leftarrow$ the set of nodes adjacent to v.
while $\overline{V} \neq \emptyset$ **do**
  Select $v \in \overline{V}$ with highest degree in the $G[U]$.
end while

While the previous heuristic assign colors to one node at a time, there are others that try first to find all nodes that can receive the same color. The set of nodes that can be colored with the same color is known as a *color class*. An algorithm based on this idea is the *largest first recursive*, due to [26]. In order to create a new color class, the heuristic proceeds as described in Algorithm 8, where $G[U]$ denotes the graph G induced by the nodes U.

A modification to the RLF procedure, called *RLF with exchange* (XRLF), was introduced in [20]. Their method consists of creating several candidate classes for the same color. For each of these classes, calculate the number of edges induced by the graph formed by the remaining nodes. XRLF is a basic heuristic in the implementation of GRASP for the coloring problem.

8.4.3 Related Work

The frequency assignment problem has been studied for the last decades under several aspects, and many papers have been published on algorithms that may be used to find exact or approximate solutions. The first methods used for the solution of the problem were informal and manually executed [19]. The main technique used to avoid the interference caused by multiple transmitters was the assignment of channels distant enough in the frequency spectrum [18]. However, this method has the undesired effect of compromising a precious area of the available spectrum, making the practical use of the method unfeasible as the number of transmitters increase and bandwidth becomes more scarce.

Since the second world war, there has been an increased use of the frequency spectrum, in large part due to military use, but also due to growing use for commercial telecommunication purposes. Since then, a discussion started on a more rational use of the frequency spectrum [21]. Applications such as TV and radio transmissions, satellites, amateur radio, and naval communications, among others, started to occupy radio frequencies in ever-growing quantities. Automatic methods started to be researched through the study of algorithms for the problem. However, mathematical models for the study of interference in radio transmissions started to be developed only during the 60s [14, 16].

In a seminal work in the area of frequency assignment, Hale proposed a mathematical model based on graph theory, which could be used in a large class of frequency assignment problems [18]. Among the contributions on this paper, there was an attempt to standardize the terminology used for the different types of problems associated with frequency assignment. It was shown, by reduction from the graph coloring problem, that the frequency assignment problem was NP-hard. It was proved that problems with distance and frequency restrictions can be reduced to a version of the problem with restrictions only on the used frequencies. Finally, the paper [18] has demonstrated the existence of subproblems of the FAP that could be solved in polynomial time.

The same work introduced a new kind of generalized graph coloring problem, known as the T-coloring problem. This is a version of graph coloring that is more adequate as a model of the frequency assignment problem. Starting in the 80s, part of the efforts for solving the FAP has been concentrated in the study of the T-coloring problem. Researchers have developed many results in this direction, including [10, 11, 27, 34]. According to these, the frequency assignment problem has been studied as a coloring problem where, for each node in the graph, there is a list of available channels.

With the fast development of mobile telephony systems, it also happened the beginning of the search for better solutions to problems associated with frequency assignment on cellular networks. This was necessary, because a particular feature of cellular systems is the predefined and fixed number of available channels. Therefore, instead of trying to minimize the number of used channels, the objective has become to produce valid assignments in such a way that the total associated interference is minimized.

8.4.4 Additional Resources

We consider the different algorithmic methods presented in the literature for the solution of frequency assignment problems. However, we give greater emphasis to methods that have commercial use. For more detailed information concerning other aspects of the literature on frequency assignment, we refer the interested reader to [31], where the authors discuss several of the classical techniques used for the solution of frequency assignment problems.

Another reference work is [24], where an analysis is made of several schemes for allocation of frequency for mobile telephony. Finally, one can cite as a recommendation the work of the CALMA project (Combinatorial ALgorithms for Military Applications). This project was created by the EUCLID (EUropean Cooperation for the Long term In Defense) to study problems related to the assignment of frequencies for telecommunications. This project had the participation of several European universities, which studied the problem using different techniques, such as Tabu Search [5], Genetic Algorithms [23], and Constraint Satisfaction [35, 36].

From the theoretical aspect, the solution methods for the frequency assignment problem can be divided into two classes: exact and approximate. Section 8.4.5 is dedicated to the presentation of exact methods for the problem. Later, in Sect. 8.5, we will present a few approximate algorithms for the frequency assignment problem.

8.4.5 Exact Algorithms

In combinatorial optimization, methods that search for exact solutions are frequently associated with mathematical programming on integer variables, or simply *integer programming*. Using integer programming, one can really model any combinatorial optimization problem; in particular, problems on graph optimization frequently use these techniques [7].

As most of the problems in integer programming require exponential time algorithms for its solution (unless $P = NP$), a few alternative solution methods have been developed, such as branch-and-cut, branch-and-bound, column generation, and polyhedral methods, among others. One should keep in mind, however, that for large-scale problems, these methods frequently present exponential time behavior, degenerating to total enumeration in many cases. Therefore, for many of the real-world instances, specially the large ones, solutions via integer programming have little practical use. Some of the application of exact methods for the frequency assignment problem will be discussed in the next sections.

8.4.5.1 Integer Programming Formulations

In an integer programming formulation, the relationship between quantities of the problem is modeled by variables, which can receive only integer values. An important special case occurs when variables in the problem can be assigned only the values 0 or 1. In this case, the resulting model is classified as a 0-1 programming model, and the variables used are called *decision variables*. Suppose that the set of antennas is given by the set $V = \{v_1, \ldots, v_n\}$ $C = \{1, \ldots, n\}$. We can define a decision variable x_{if} such that its value is 1 when channel f is assigned to antenna i, and 0 otherwise. In mathematical notation:

$$x_{if} = \begin{cases} 1 \text{ if freq}(i) = f \\ 0 \text{ otherwise.} \end{cases}$$

Another decision variable can be created to represent if a frequency is in use or not:

$$y_f = \begin{cases} 1 \text{ if frequency } f \text{ is in use} \\ 0 \text{ otherwise.} \end{cases}$$

Aardal et al. [1] proposed an integer formulation for the FAP using these decision variables. Their formulation can be presented as follows:

(PAFVP)

$$\min \quad \sum_{f \in C} y_f \tag{8.3}$$

$$\textbf{s.t.} \quad \sum_{f} x_{i,f} = 1 \quad \forall i \in V \tag{8.4}$$

$$x_{i,f} + x_{j,g} \leq 1 \; \forall i, j \in V, \forall f, g \in C, \; |f - g| < d(i, j) \tag{8.5}$$

$$x_{i,j} \in \{0, 1\} \quad \forall i \in V \tag{8.6}$$

In the FAPVP formulation, the objective function (8.3) represents the sum of used frequencies. This allows one to minimize the total number of frequencies used in the assignment (span). Equations (8.4) are used to guarantee that each antenna will receive one and only one channel. The restriction therefore disallows any unfeasible solutions with zero or more than one assigned channels. The set of inequalities (8.5) represents the restriction that says that, for any two antennas that cause interference, only one of them can receive the specified frequency. This group of inequalities, therefore, can be used to model the interference restrictions. Finally, in the set of Eqs. (8.6), we have the integrity restrictions that define the decision variables used in the model.

Branch-and-cut is a technique frequently used to solve integer programming models. The basic strategy used by branch-and-cut is solving a simpler version of the original problem and use this result to derive a complete solution. The main difficulty encountered with branch-and-cut is that it can degenerate into a complete enumeration of the solution space. In [1], this method is used to provide solutions for the FAPVP formulation, displayed above. The first step in the application of branch-and-cut on the algorithm proposed in [1] is to perform preprocessing in the FAP instance. The goal is to remove redundancies that might exist in the original formulation. With this, the number of variables may be reduced considerably, depending on the structure of the instance.

The establishment of lower bounds is another factor that has an impact in the capacity of finding good solutions for the FAP. This can be done through the inclusion of structural information, such as the size or large cliques in the interference graph. With the use of good lower bounds, the branch-and-cut algorithm can decide more easily if a solution is good or not for the given instance.

Based on the formulation given above, which has the objective of minimizing the total number of used channels, one can derive a integer programming model in which the objective is to minimize the total interference occurring in the system. In order to do this, it is necessary to add a few constants $c_{i,j} \in [0, 1] \forall i, j \in V$ representing the value of the co-channel interference between each pair of antennas. Similarly, we define $a_{i,j} \in [0, 1] \forall i, j \in V$, which represents the adjacent-channel interference in the system. Then, we have the following formulation:

(PAFC)

$$\min \sum_{i,j,f} x_{i,f}x_{j,f}c_{i,j} + \sum_{i,j,f} x_{i,f}x_{j,(f+1)}a_{i,j} + \sum_{i,j,f} x_{i,f}x_{j,(f-1)}a_{i,j} \quad (8.7)$$

such that

$$\sum_{f} x_{i,f} = 1 \quad \forall i \in V \quad (8.8)$$

$$x_{i,f} + x_{j,g} \leq 1 \quad \forall i, j \in V, \forall f, g \in C, \; |f - g| < a(i, j) \quad (8.9)$$

$$x_{i,j} \in \{0, 1\} \quad \forall i \in V \quad (8.10)$$

In this formulation, the objective function represents the summation of co-channel and adjacent-channel interference. The multiplication of variables in the summation guarantees that the factors will be used only when both variables are equal to 1. However, with this change, the objective function loses its linearity, becoming a sum of quadratic terms. This makes more difficult to solve such a formulation using conventional methods, such as linear relaxations.

8.5 Aproximate Algorithms

Given the intrinsic difficulty of solving instance of the frequency assignment problem, several approximate algorithms have been proposed with the goal of offering a near-optimal solution and with relatively low computational effort. A few of these approximate algorithms also offer a guarantee of performance, based on the distance to the optimum. In Sect. 8.5.1, we present an algorithm with these features. Many heuristics for the FAP are based in existing heuristics for the graph coloring problem. In this section, we will present some of these heuristics that seem to be most effective for the frequency assignment problem.

8.5.1 Algorithms with Approximation Guarantee

Some non-exact algorithms offer an approximation guarantee, which is usually provided in relation to the optimal value of the solution. In formal terms, we are given an approximation algorithm A to an minimization problem Π, and a rational

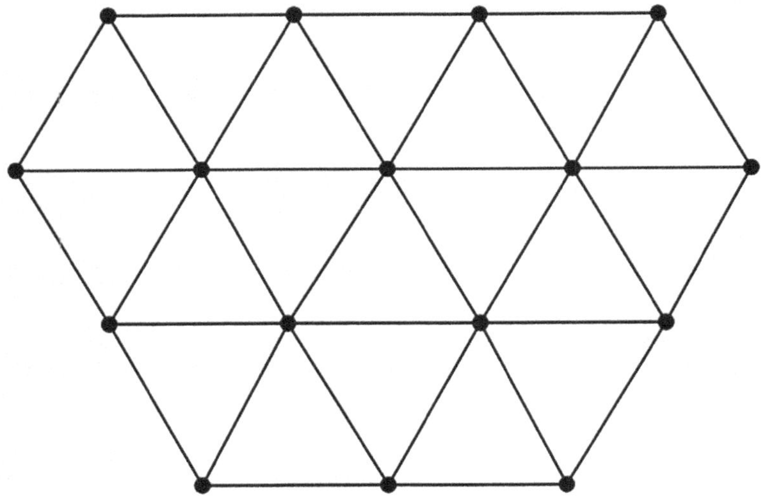

Fig. 8.2 A triangular reticulate graph

number $\epsilon > 1$. We say that algorithm A has approximation guarantee ϵ (or that A is an ϵ-approximation algorithm) if for each solution $s \in A(\Pi)$ returned we have

$$f(s) \leq \varepsilon \cdot f(s^*)$$

where $s^* \in A(\Pi)$ represents an optimal solution for the problem. In the case of maximization problems, a similar relation is true, with an inversion in the sign of the inequality. That is, given a value $\epsilon < 1$, we have

$$f(s) \geq \varepsilon \cdot f(s^*).$$

It can be proved that, unless $P = NP$, there are no algorithms with constant approximation guarantee for the FAP [15]. Despite this negative result, there is nothing to prevent the development of approximate algorithms for simplified versions of the problem. This was the path explored in [30]. In this work, an approximate algorithm was developed for the graph coloring problem in a triangular reticulate (see Fig. 8.2), where each node has an assigned demand. Another restriction imposed in this problem is that minimum separation function be of the following form:

$$d(i, j) = \begin{cases} 1 \text{ if } (i, j) \in E(G) \\ 0 \text{ otherwise.} \end{cases}$$

For this restricted version of the frequency assignment problem, one can prove the following result:

Teo 8.1 ([30]) *Given a reticulated graph G and a demand vector w for the nodes of G, there is a ε-approximate algorithm Π that run in polynomial time and such that*

$$f(\Pi(G, w)) \leq \varepsilon \cdot f(s^*),$$

where s^* is the optimal solution for the problem, and $\varepsilon \geq 4/3$.

8.5.2 Heuristics

A second class of non-exact algorithms for combinatorial optimization problems is composed by heuristics, which offer no guarantee of approximation. We present in the following sections some of the best known heuristics among the several that have been proposed for the frequency assignment problem.

8.5.2.1 T-Coloration

The T-Coloration heuristic received its name, because it was originally used in the context of the T-Coloring problem [10]. The main objective of the heuristic is to find a feasible solution; it toward this goal, it employs the notion of "saturation" and "spacing degree."

The saturation of a node is the number of colors available for it in the moment. This idea was first used in the DSATUR algorithm, proposed in [6, 33] (see Sect. 8.4.2). The spacing degree represents the number of nodes that are neighbors of the current node and that have no assigned colors. Using these two measures, the T-coloration algorithm tries to find the next node to which a valid color can be assigned.

Using this algorithm for the solution of the frequency assignment problem, it is necessary to calculate, at each iteration, the values of saturation and spacing degree. Algorithm 9 shows a pseudo-code that gives an idea of how the T-Coloring heuristic works.

8.5.2.2 Dual Greedy Algorithm

The dual greedy algorithm received this name because, despite using a greedy strategy, its selection is based on dual values for the problem. It also avoids taking an intermediate decision whenever possible. The idea for the algorithm has been used for other problems, such as location in electrical circuits [22]. First, the algorithm tries to characterize what could be a bad choice for the pair (antenna, frequency channel). In order to do this, it searches, among all valid combinations of these two values, the ones that have highest value of total interference, and then discarding that value. This process is repeated until only one antenna with its associated

Algorithm 9 T-coloration heuristic

```
Initialize sat[i] ← 0, ∀i ∈ {1, ..., |V|}.
Initialize esp[i] ← 0, ∀i ∈ {1, ..., |V|}.
Let U = V.
while U ≠ ∅ do
    Select v_i ∈ U with largest combined value of sat[i] and esp[i].
    Let U ← U − {v_i}
    if there are available channels for v_i then
        Let A_i ← set of available channels for v_i.
        Randomly assign to v_i a channel from A_i.
    else
        Assign a random channel to v_i.
    end if
    // Update values
    for each j ∈ {1, ..., |V|} do
        Let sat[j] ← number of channels that v_j cannot use.
        Let esp[j] ← number of neighbors of v_j with no assigned color.
    end for
end while
```

Algorithm 10 Dual greedy heuristic

```
Initialize U to be the set of all pairs (v, f) such that v ∈ V and f ∈ C.
Let pena[i,j] ← 0 ∀i ∈ {1, ..., |V|} and ∀j ∈ {1, ..., |C|}
while U ≠ ∅ do
    for all pairs (v, f) ∈ U do
        if (v, f) violated any restriction or (v, f) causes co-channel or adjacent-channel
        interference then
            pena[v, f] ← pena[v, f] + 1
        end if
    end for
    Select the pair (v, f) with lowest value of pena[v,f] (break ties randomly)
    Assign channel f to antenna v
    Let U ← U − (v, f)
end while
```

frequency channel is left. These steps are repeated, while there are antennas without an assigned frequency channel. In Algorithm 10, we have a pseudo-code description for the heuristic, such as presented by Borndörfer et al. [3].

It is easy to see that the dual greedy algorithm has a performance that is worse than that of the T-Coloring heuristic. At each iteration, all valid pairs of (antenna,frequency channel) need to be checked for assignment to a new node. This makes this algorithm to have worse performance in comparison to similar algorithms. According to Borndörfer et al. [3], the most important factor in the success of this algorithm is a good choice of a function that measures the badness of a specific antenna-frequency pair. Using Fibonacci Heaps [9] to keep the set of pairs, one can obtain computational complexity similar to $O(|C|^2|V|\log(|C||V|) + |C|^2|E|)$, with space complexity equal to $O(|C||V| + |E|)$.

8.5.2.3 Minimum Cost Flows

The *minimum cost flow* heuristic, discussed in [3], was created with the goal of equalizing the use of different frequency channels. A model based on the minimum cost flow problem is used by the algorithm with this purpose.

The minimum cost flow heuristic tries to modify a frequency assignment y into an assignment y' while maintaining intact the property that, $\forall(v, w) \in E, v < w$, one of the following conditions is valid:

$$\begin{cases} y(v) \leq y(w) \Rightarrow y'(v) \leq y'(w) \\ \qquad\qquad \bigwedge \\ y(v) > y(w) \Rightarrow y'(v) \geq y'(w) \end{cases}$$

The objective of these conditions is to produce a frequency assignment that utilizes the largest number of available channels, making the usage of channels among antennas more even. A way of satisfying these conditions consists of solving a minimum cost flow problem, which has as its base the graph of interferences G. Functions d, c^{co} e c^{ad} are used to compute the cost coefficients and the arc capacities. To use this heuristic, it is necessary to solve the minimum cost flow problem, which can be done using well-known method such as the simplex algorithm. The result obtained in this say is a frequency assignment with an improved distribution of channels among transmitters. This usually leads to lower interference.

According to results reported in [3], this heuristic doesn't offer substantial gains in comparison to other techniques discussed. It also suffers from low computational performance, due to the necessity of solving a linear programming problem. For this reason, MCF is not frequently used as an implementation strategy for the frequency assignment problem.

8.5.3 *Metaheuristics*

In this section, we present the main results for the frequency assignment problem that have been developed through the use of metaheuristics.

8.5.3.1 Tabu Search

Tabu search has been used by some researchers to find high-quality solutions for the frequency assignment problem. For example, in [2], a tabu search algorithm was proposed to the frequency assignment problem applied to cellular telephony. The methodology used there is to perform frequency assignment on blocks of channels, in such a way that will maintain the minimum distance between antennas that member of the same base station. The movements of neighborhood search consist of exchanging a whole block of channels by another, with the goal of minimizing the total interference occurring in the system.

Another kind of movement used by the tabu search algorithm consists of exchanging all channels used inside a cell for channels that can generate less interference. This process is repeated until all the demand in the cell covered by that base station is satisfied. Experiments performed with this algorithm used two instances of cell management from Bell Mobilité. However, the tests performed in the paper involved only the comparison with a search local algorithm, which does not provide enough information to compare it in relation to other metaheuristic methods.

8.5.3.2 Genetic Algorithms

Genetic algorithms have been frequently used as a method for solving the frequency assignment problem. In [12], we are presented with an algorithm genetic for the FAP that simultaneously considers multiple objectives. Each objective is codified as multiplicative factors applied to the objective function. In their implementation, two representations of the problem were used. In representation $R1$, the size of each chromosome is n, the number of transmitters. Each element is the number of the channel assigned to that transmitter.

In representation $R2$, the transmitters that receive the same signal are grouped sequentially, ordered by increasing channel numbers. According to performed experiments, representation $R2$ displayed superior performance in the tests. In the same work, a simple form of parallel genetic algorithm is presented. In this version of the GA, several processors execute the algorithm, while there is occasional exchange among the populations of chromosomes with very small objective function values.

8.5.3.3 GRASP

The GRASP (General Randomized Adaptive Search Procedure) metaheuristic was also used to provide near-optimal solutions for the frequency assignment problem. For example, in [28], we have a GRASP for the FAP that is similar to the algorithm developed later in this book. The following differences are notable, however:

- The objective function minimized by the GRASP in [28] is concerned with violations of minimum separation between channels. In the GRASP developed later in this book, the goal is consider not only this value but also the total interference in the system caused by co-channel and adjacent-channel interferences (similar to what is proposed in [3].
- The method used in the article cited above does not consider the specialized structure of the frequency assignment problem when applied to cellular telephony. In

the algorithms presented later, however, we will consider transmitters grouped in a set of base stations, which allows for more efficient solutions.

- Finally, the GRASP algorithm implemented in [28] differs from the one provide later in this book. For example, the local search described in that article is based on simulated annealing, which we will use a greedy heuristic.

Due to these difference between the approaches of [28] and the one presented in the next chapter, it is hard to directly compare the results of the respective GRASP implementations.

8.6 Alternative Formulations

A current tendency, viewed in several papers on the frequency assignment problem, is the definition of new models to which new strategies of frequency assignment can be used with better results. One can cite, for example, [8], where a new mathematical mode, incorporating cumulative interference is introduced. A second example can be seen in [13].

In that work, the restrictions that are part of the frequency assignment problem are divided into two types: hard restrictions and soft restrictions. This has allowed the authors to use a constraint satisfaction method, also used in [32]. In [4], an alternative model for the frequency assignment based on integer programming is used. The model takes into consideration the power associated with transmitters as a determining fact of the interference.

8.7 Conclusion

In this chapter, we presented several computational methods used for the solution of frequency assignment problems. All these methods have been developed for the most general version of the problem, and many do not take into consideration the unique characteristics of the FAP for cellular telephony (even when some of the discussed algorithms have been used with this objective).

While the presented algorithms are very different, their main disadvantage is not being adapted to some of the important characteristics of cellular systems, such as the location of transmitters in the base station and fixed local interference, among others.

In the next chapter, we present an improved solution for the frequency assignment problem on cellular networks. We will use the metaheuristic methodology to speed up the solution process.

References

1. K.I. Aardal, A. Hipolito, C.P.M. Van Hoesel, B. Jansen, C. Roos, T. Terlaky, A branch-and-cut algorithm for the frequency assignment problem. Technical report, Eindhoven University, 1998
2. P.M. Adjakpl'e, B. Jumard, *Greedy and Tabu Search Heuristics for Channel Block Assignment in Cellular Systems* (Les Cahiers du GERAD, 1997)
3. R. Borndörfer, A. Eisenblätter, M. Grötschel, A. Martin, Frequency assignment in cellular phone networks. Technical report, Konrad-Zuze-Zentrum für Informationstechnik, Berlin, 1997
4. R. Borndörfer, A. Eisenblätter, M. Grötschel, A. Martin, The orientation model for frequency assignment problems. Technical report, Konrad-Zuse-Zentrum für Informationstechnik Berlin, 1999
5. A. Bouju, J. Boyce, C. Dimitropoulos, G. vom Scheidt, J. Taylor, Tabu search for the radio links frequency assignment problem, in *Applied Decision Technologies* (UNICOM, London, 1995)
6. D. Brélaz, New methods to color the vertices of a graph. Commun. ACM **22**(4), 251–256 (1979)
7. R.E. Campelo, N. Maculan, *Algoritmos e heurísticas: desenvolvimento e análise de performance* (Eduff/Furnas centrais elétricas, 1994)
8. A. Capone, M. Trubian, Channel assignment problem in cellular systems: a new model and a tabu search algorithm. IEEE Trans. Veh. Technol. **48**(4), 1252–1260 (1999)
9. T.H. Cormen, C.E. Leiserson, R.L. Rivest, *Introduction to Algorithms* (MIT Press, Cambridge, 1990)
10. M.B. Cozzens, F.S. Roberts, T-colorings of graphs and the channel assignment problem. Congressum Numerantium **35**, 191–208 (1982)
11. M.B. Cozzens, F.S. Roberts, Greedy algorithms for T-colorings of complete graphs and the meaningfulness of conclusions about them. J. Comb. Inform. Syst. Sci. **16**(4), 286–299 (1991)
12. W. Crompton, S. Hurley, N.M. Stephens, A parallel genetic algorithm for frequency assignment problems, in *IEEE International Symposium on Signal Processing, Robotics and Neural Networks*, Lille (IMACS/IEEE, 1994), pp. 81–84
13. N. Dunkin, J. Bater, P.J., D. Cohen, Towards high order constraint representations for the frequency assignment problem. Technical report, Department of Computer Science Royal Holloway College, University of London, 1998
14. H. Eden, H.W. Fastert, K.H. Kaltbeitzer, More recent methods of television network planning and the results obtained. E.B.U. Rev **60-A**, 54–59 (1960)
15. A. Eisenblätter, A frequency assignment problem in cellular phone networks (extended abstract), in *Network Design, Connectivity and Facility Location*, volume 35 of *Dimacs Series in Discrete Mathematics and Theoretical Computer Science* (American Mathematical Society, 1997)
16. H.W. Fastert, The mathematical theory underlying the planning of transmitter networks. E.B.U. Rev **60-A**, 60–69 (1960)
17. M.R. Garey, D.S. Johnson, *Computers and Intractability—a Guide to the Theory of NP-Completeness* (W. H. Freeman and Company, 1979)
18. W.K. Hale, Frequency assignment: theory and applications. Proc. IEEE **68**(12), 1497–1513 (1980)
19. D.M. Jansky, *Spectrum Management Techniques* (Don White Consultants, 1977) Citado em [W.K. Hale, Frequency assignment: theory and applications. Proc. IEEE **68**(12), 1497–1513 (1980)]
20. D.S. Jonhson, C.A. Aragon, L.A. Mcgeoch, C. Schevon, Optimization by simulated annealing: an experimental evaluation—Part II (graph coloring and number partition). Oper. Res. **31**, 378–406 (1991)
21. JTAC, *Spectrum Engineering—the Key to Progress* (IEEE, New Yor, 1968). Citado em [W.K. Hale, Frequency assignment: theory and applications. Proc. IEEE **68**(12), 1497–1513 (1980)]

22. M. Jünger, A. Martin, G. Reinelt, R. Weismantel, Quadratic 0/1 optimization and a decomposition approach for the placement of eletronic circuits. Math. Program. **63**, 257–279 (1994)
23. A. Kapsalis, V.J. Rayward-Smith, G.D. Smith, Using genetic algorithms to solve the radio link frequency assignment problem, in *Proceedings 2nd Int. Conf. on Artificial Neural Networks and Genetic Algorithms*, ed. by D.W. Pearson, N.C. Steele, R.F. Albrecht, Ales (Springer, Berlin, 1995), pp. 37–40
24. I. Katzela, M. Naghshineh, Channel assignment schemes for cellular mobile telecommunication systems. IEEE Pers. Commun. Mag. **3**(3), 10–31 (1996)
25. M. Laguna, R. Martí, A GRASP for coloring sparse graphs. Technical report, Graduate School of Business, University of Colorado, Boulder, CO 80309-0419 USA, 1998
26. F.T. Leighton, A graph coloring algorithm for large scheduling problems. J. Res. Nat. Bur. Standard **84**(6), 489–506 (1979)
27. D.D. Liu, T-colorings of graphs. Discrete Math. **101**, 203–212 (1992)
28. X. Liu, P.M. Pardalos, S. Rajasekaran, M.G.C. Resende, A GRASP for frequency assignment in mobile radio networks. DIMACS Series Discrete Math. Theor. Comput. Sci. **52**, 195–201 (2000)
29. D.W. Matula, G. Marble, J.D. Isaacson, *Graph Theory and Computing*, chapter Graph Coloring Algorithms, ed. by R.C. Read (Academic Press, New York, 1972), pp. 104–122
30. C. McDiarmid, B. Reed, Channel assignment and weighted colouring (1997)
31. R.A. Murphey, P.M. Pardalos, M.G.C. Resende, Frequency assignment problems, in *Handbook of Combinatorial Optimization*, ed. by D.-Z. Du, P. M. Pardalos, volume 4 (Kluwer Academic Publishers, Dordrecht, 1999), pp. 295–377
32. G. Ottosson, M. Carlsson, Using global constraints for frequency allocation. Technical Report TR-97-07, ASTEC, 1997 Citado em [N. Dunkin, J. Bater, P.J., D. Cohen, Towards high order constraint representations for the frequency assignment problem. Technical report, Department of Computer Science Royal Holloway College, University of London, 1998]
33. J. Peemöller, A correction to Brelaz's modification of Brown's coloring algorithm. Commun. ACM **26**(8), 595–599 (1983)
34. F.S. Roberts, T-colorings and graphs—recent results and open problems. Discrete Math. **93**, 229–245 (1991)
35. T. Schiex, H. Fargier, G. Verfaillie, Valuated constraint satisfaction problems; hard and easy problems, in *14th International Joint Conference on Artificial Intelligence* (1995)
36. G. vom Scheid, Extension of genet to n-ary partial constraint satisfaction problems and constraint satisfaction optimisation problems. Master's Thesis, King's College London, 1997
37. D.J.A Welsh, M.B. Powell, An upper bound for the chromatic number of a graph and its application to timetabling problems. Comput. J. **10**, 85–86 (1967)

Chapter 9
The Metaheuristic Strategy for AI Search and Optimization

Carlos A. S. Oliveira

Abstract The design of exact algorithms running in polynomial time for problems belonging to the class NP-hard has been one of the main open problems in the area of theoretical computer science. For this reason, when one needs to solve instances of these problems that are sufficiently large, it is necessary to employ methods that may result only in an approximate solution. This chapter discusses metaheuristic strategies that are commonly used in the solution of hard combinatorial optimization problems. Approximation algorithms with performance guarantees offer a powerful approach to tackling NP-hard problems, providing solutions that are provably close to optimal. However, the design of such algorithms relies heavily on the specific properties of the problem and may not be generalizable across different problem domains. For problems where no approximation guarantee is feasible, heuristic methods remain an essential tool in the practitioner's arsenal.

Keywords Metaheuristics · Math programming · AI algorithms · Optimization

9.1 Approximation Algorithms

Approximate solutions for NP-hard problems can be broadly classified into two categories: those with a guarantee of approximation performance and those without such a guarantee. The first category includes algorithms that provide a solution within a specified distance from the global optimum. This distance is typically measured as a multiplicative factor of the optimal value, and these algorithms are known as **approximation algorithms** with performance guarantees.

C. A. S. Oliveira (✉)
Principal Inventive Scientist, Network Chief Technology Organization, AT&T Labs Inc., Bellevue, WA, USA
e-mail: carlos.oliveira@att.com

C. A. S. Oliveira, M. P. Pardalos (eds.), *Handbook of Artificial Intelligence and Data Sciences for Routing Problems*, Springer Optimization and Its Applications 219, https://doi.org/10.1007/978-3-031-78262-6_9

Given a constant $\varepsilon > 1$, an ε-approximate algorithm for a problem Π is one that, for any instance I_Π of Π, returns a solution s such that:

$$f(s) \leq \varepsilon \cdot f(s^*),$$

where $f(s)$ is the cost or value of the solution produced by the algorithm, and $f(s^*)$ is the cost or value of the optimal solution s^* to the instance I_Π. The parameter ε represents the approximation ratio, and it quantifies how close the algorithm's solution is to the optimal solution.

9.1.1 Properties and Techniques in Approximation Algorithms

Algorithms with performance guarantees typically exploit specific properties of the problem at hand. These properties may include structural characteristics, combinatorial features, or problem-specific inequalities that help in bounding the difference between the approximate and optimal solutions. Common techniques used in the design of approximation algorithms include:

- **Greedy Algorithms**: Greedy methods build a solution step by step, choosing the locally optimal option at each stage. For certain problems, like the classic *Knapsack Problem* or *Set Cover*, greedy algorithms can offer provable approximation guarantees. For example, the greedy algorithm for the Set Cover problem provides a solution that is within a factor of $\log n$ of the optimal solution, where n is the number of elements to be covered.
- **Relaxation and Rounding**: This approach involves relaxing an NP-hard problem into a more tractable form (e.g., relaxing an integer linear program into a linear program) and then rounding the solution of the relaxed problem to obtain a feasible solution to the original problem. Techniques such as *linear programming (LP) relaxation* and *semidefinite programming (SDP) relaxation* are widely used in designing approximation algorithms for problems like *Vertex Cover* and *MAX-CUT*.
- **Dynamic Programming with Approximation**: In some cases, dynamic programming can be combined with approximation techniques to achieve near-optimal solutions. An example of this is the *Fully Polynomial-Time Approximation Scheme (FPTAS)* for the Knapsack Problem, which provides a solution arbitrarily close to the optimal one, with a trade-off between accuracy and computation time.
- **Local Search and Heuristics**: Local search methods start with an initial solution and iteratively improve it by exploring neighboring solutions. While local search does not always guarantee an optimal solution, for some problems, it can be shown to provide good approximate solutions within a certain bound.

9.1.2 Challenges in Approximation

Despite the success of approximation algorithms in various domains, designing such algorithms is challenging for many NP-hard problems. Some problems do not admit any approximation guarantees unless $P = NP$, as indicated by results from complexity theory. For instance, problems like *Exact Cover* or *Clique* are known to be hard to approximate within any reasonable factor unless significant breakthroughs are made in complexity theory.

In fact, a cornerstone result in complexity theory is the **inapproximability** of certain problems. For example, it is known that there is no polynomial-time approximation algorithm for the *Travelling Salesman Problem (TSP)* with general distances that guarantees a solution better than any constant factor unless $P = NP$ [10]. Similar hardness results exist for other problems such as *Set Packing* and *Submodular Function Maximization*.

9.1.3 Without Approximation Guarantees

On the other hand, the second category includes **heuristic algorithms** that do not provide any formal guarantee on the quality of the solution but may perform well in practice. Examples include *metaheuristic algorithms* like *genetic algorithms*, *simulated annealing*, and *ant colony optimization*. These algorithms are often used when obtaining even a guaranteed approximate solution is computationally infeasible. Although they lack theoretical guarantees, they are widely used for their flexibility and effectiveness on large, complex instances.

Among approximate algorithms without approximation guarantee, we will find many heuristic methods. A heuristic is an algorithm that uses specific knowledge about a problem one wishes to solve with the goal of finding quickly, and with high probability, a solution that is good enough for the required purposes. The development of heuristics for combinatorial optimization problems is a very active research area.

Despite of the advantages possessed by heuristic methods in terms of speed and average quality of solution, certain problems make its use difficult. For example, it is known that the quality of the solutions produced by a heuristic algorithm for optimization can vary, according to the features of the problem instance that needs to be optimized. Therefore, a heuristic can be efficient for a few instances but at the same time have a non-satisfactory behavior when used on other equally important instances of the problem. The topology of an instance as well as its internal structure is characteristics that can reduce the performance of heuristics.

Another difficult problem that happens in the use of heuristics for the solution of combinatorial optimization problems is the tendency they have to be trapped into local optima. A local optimum is a point s^* in the solution space such that, for each point s' in the neighborhood of s^*, we have $f(s') \geq f(s^*)$. In other words, any

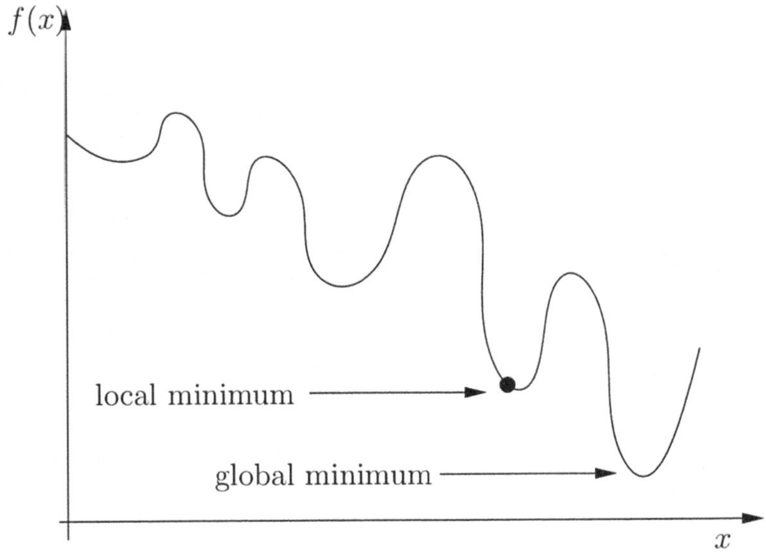

Fig. 9.1 Representation of local minima and global minimum of a function

element in the neighborhood of s^* has a corresponding objective value that is the same or worse than $f(s^*)$. Figure 9.1 shows an example of continuous, non-convex function with several local minima. Greedy heuristics, which always try to improve the objective function, can get easily trapped into these points without the ability to achieve a global optimum.

As a way of solving some of the perceived problems found in heuristic search procedures, a number of more generalized heuristics have been proposed. They are known as metaheuristics, and a number of features mark the difference between them and more traditional heuristic methods. First, metaheuristics are general. Due to their general framework, each metaheuristic can be implemented for a large number of problems. This is possible, because the general description of a metaheuristic has no reference to particular problems. These generic strategies can be adapted as needed to solve a specific combinatorial problem. The metaheuristic only provides common ideas and techniques that need to be present in any implementation.

Frequently, heuristics are used in several distinct problem domains while changing only some of the input parameters. Such parameters are used to calibrate the procedure in such a way that the solutions produced will satisfy the requirement of a specific problem.

Another important characteristic of metaheuristics is the use of randomization techniques. Through the use of random choices, metaheuristics reduce the possibility of becoming trapped in a local optimum during the search process. In this way, metaheuristics try to eliminate or at least ameliorate the main problem encountered in heuristics that use local search. The application of randomized techniques also favors a better exploration of the solution space, allowing an improved search time in some or even most of the cases.

Yet another important feature of metaheuristics is related to its behavior during the solution of large-scale combinatorial optimization problems. Due to the discussed features, metaheuristics are capable of finding good solutions even for very large instances of NP-hard problems. Many of these problems cannot be tackled using exact methods, such as integer programming. Moreover, for some problems for which heuristics have not been previously developed, metaheuristics remain as one of the few good alternatives for finding solutions.

Several metaheuristics have been proposed in the literature, with the objective of exploring different search mechanisms. Among the heuristics that have been demonstrated as promising approaches for general use, one can cite Tabu Search [12, 13], Simulated Annealing [22], Genetic Algorithms [18], and GRASP (Greedy Randomized Adaptive Search Procedure) [6].

In the next few sections, we will discuss three of these metaheuristics. In Sect. 9.2, we present the Tabu Search method. Section 9.3 is dedicated to the Simulated Annealing metaheuristic. Finally, Genetic Algorithms are presented in Sect. 9.4. GRASP will be discussed at the end of the chapter.

9.2 Tabu Search Metaheuristic

Tabu Search is a global search method used in combinatorial optimization and introduced by Glover [12, 13]. The process used by tabu Search tries to avoid the tendency of heuristics of becoming trapped into local minima. To escape such local minima, the tabu search metaheuristic uses a set of operations, such as storing solution that were previously found in a tabu list. The tabu list is a data structure with t positions that allows the algorithm to "remember" features from the last t solutions visited by the algorithm. The avoid situations in which the algorithm might return to solutions that were previously explored. The size of the tabu list is one of the parameters that can be adjusted in the metaheurist; its value needs to be empirically determined.

According to Glover [12], tabu search aggressively explores the solution space of a problem, searching for a better configuration at each movement. The algorithm starts from a solution s and searches the neighborhood of s for other solutions with the best objective function among the elements of the neighborhood. In case this new element s' has cost that is less than the cost of the original solution, the algorithm will update s to refer to s', and s' is immediately added to the tabu list. If there is no such solution s' such that $f(s') < f(s)$, however, the s' will be accepted only if it is not currently stored in the tabu list. The algorithm then will restart its search using the new values for s and the elements stored in the tabu list.

Another important component in the tabu search heuristic is its use of an aspiration criterion. When all possible choices for new solutions are already

Algorithm 11 Tabu search metaheuristic

s is the initial solution
// Init the tabu list
$L \leftarrow \emptyset$.
while iter < max_iter **do**
  Let s be the current solution. Find the neighbor s' with smallest value of objective function $f(s')$.
  if $s' \notin L$ **and** $f(s') \leq f(s)$ **then**
    Let $L \leftarrow L \cup \{s\}$
    Let $s \leftarrow s'$
  else
    if $f(s') \leq Asp(f(s))$ **then**
      // Update the aspiration criterion
      $Asp(f(s')) \leftarrow f(s) - \Delta f$
      Let $L \leftarrow L \cup \{s\}$
      Let $s \leftarrow s'$
    end if
  end if
  Remove the oldest element from L.
  Let max_iter $\leftarrow$ max_iter + 1
end while

contained in the tabu list, an aspiration criterion can offer a way for the algorithm to return one of the elements on the tabu list. There are several ways of defining the aspiration criteria. A possible way would be to require that a solution that is part of the tabu list be revisited only if it leads to a solution that is better than the current solution. A pseudo-code showing some of the main features of the tabu search metaheuristic is displayed on Algorithm 11.

The tabu search metaheuristic has been used for the solution of a large number of problems in combinatorial optimization. An exhaustive list was presented by Glover and Laguna [15]. Some recent works have the goal of parallelizing this metaheuristic. An example is given in [29].

9.3 Simulated Annealing for the FAP

The simulated annealing metaheuristic is an optimization method proposed by Kirkpatrick et al. [22]. Simulated annealing was developed with the objective of simulating the process of cooling on physical systems. When cooled, a metal alloy in liquid state may become brittle if the cooling process is too fast. This happens, because it can't find an energy state that can guarantee stability. In order for the cooling to happen at a uniform manner, it is necessary that the temperature be reduced at a slow pace, with gradual variations.

With optimization problems, one can make an analogy between the value of the objective function and the energy state of a cooling metal alloy, as described in the previous paragraph. The final goal of a search procedure is to find a configuration with the minimum value of objective function cost, in the same way that in the cooling process the goal is to find a state of minimum internal energy.

The achieve this objective, simulated annealing is executed in several phases. Each phase is performed at a lower temperature, in such a way as the cooling process is completely simulated. In each phase of the simulated cooling process, several iterations are performed that try to improve the existing solution. For example, starting from a configuration s_0, the algorithm may find a new solution s_1 through a search of the neighborhood of s_0. If $f(s_1) \leq f(s_0)$, then the algorithm accepts s_1 as the new current configuration and the process continues. In this case, simulated annealing behaves exactly as a standard local search procedure (a downhill algorithm).

On the other hand, if $f(s_1) > f(s_0)$, then s_1 may still be accepted by simulated annealing, with probability given by

$$e^{-\Delta F/T},$$

where $\Delta F = f(s_0) - f(s_1)$ represents the difference between the objective function values for s_0 and s_1 and T is the current temperature, which can be freely specified by the designer of the metaheuristic in order to reflect the cooling process. The value $e^{-\Delta F/T}$ is known as the *Metropolis criterion* of acceptance. It is based on the Boltzman distribution, which is used to model the physical phenomenon of cooling.

A pseudo-code description for simulated annealing can be found on Algorithm 12. At the beginning of processing, a temperature is initialized to a very high value. After that, the search process is started using the method described in the previous paragraph. The metaheuristic continues with these iterations until the solution reaches an equilibrium. This can be indicated, for example, by the execution of a fixed number of iterations (`MAX_TRY`) without any improvement in the objective function. After the end of one of these steps, the temperature is reduced by a value ΔT, and the computation is restarted. This procedure ends when temperature reaches zero. Then, the result of the computation is simply the value of the current solution at that moment.

One can see that the simulated annealing metaheuristic is composed of several phases, where at each one of them the freedom to incorporate worse solutions is reduced. This makes it possible for the metaheuristic to escape from local optima that may happen at the beginning of the process. In the other hand, a gradual temperature reduction makes the probability $e^{-\Delta/T}$ smaller at each iteration. Therefore, in each phase, it becomes harder to move to worse solutions. This guarantees that the metaheuristic will converge to a final unique solution.

One can prove that this algorithm asymptotically approaches the global optimum with probability 1 [32]. Despite this, as it happens with other metaheuristics, the behavior of simulated annealing depends on the careful adjusting of parameters that are application-dependent. For example, parameters like T and ΔT are needed to determine the convergence criteria.

In the last years, simulated annealing has been used extensively for the solution of several problems. This metaheuristic has been applied to classical problems, such

Algorithm 12 Simulated annealing metaheuristic

```
Let T ← T_0
Initialize current solution s
repeat
    M ← 0
    repeat
        Let s be the current solution.
        Find neighbor s' ∈ N(s) with minimum objective function cost
        Let ΔF ← f(s) − f(s')
        if s' ≤ s then
            Let s ← s'
        else
            if random[0,1] < exp(−ΔF/T) then
                Let s ← s'
            end if
        end if
        Let M ← M + 1
    until M ≥ MAX_TRY
    Let T ← T − ΔT
until T <= 0
```

as the ones presented in the first lists of NP-hard problems [19–21]. It has also been used on image processing problems [2], DNA mapping[16], and experimental design [25]. Another area of interest in the design of simulated annealing is the development of parallel implementations, such as the one proposed in [1]. For additional information, one can check reference works such as [4].

9.4 Genetic Algorithms

Genetic algorithms are a class of optimization methods that try to simulate the process of natural evolution of species through natural selection. A few of the first works in this area can be seen in [18, 30, 31]. Genetic algorithms have been used with success in the last years for several combinatorial optimization problems. Contrary to tabu search and GRASP, applications also appear in areas other than combinatorial optimization, such as pattern recognition. In its simples form, a genetic algorithm encodes the solution to a problem in a binary representation, to which the name *chromosome*g is given. A chromosome represents an element on the set of feasible solutions.

A function f is defined and called the fitness function or objective. The function f is used to evaluate the desirable features of a chromosome. For example, one can establish that the smaller the value of $f(x)$, the better adapted chromosome x will be to the fitness function. The central idea of an algorithm genetic is to create an initial population of chromosome, and starting from a few operation (known as genetic operators) to derive new generations of chromosomes. The genetic operators have

0	1	1	0	0	1	0	1	1	0

Fig. 9.2 Sample chromosome for the packing problem. Each position represents the inclusion of exclusion of an object. The size of the chromosome represents the number of objects that can be included in the backpack

features that allow the chromosomes generated by them to become more adapted to the fitness function as the algorithm progresses.

As an example, one can check the application of a genetic algorithm to the packing problem (see Sect. 7.4). Initially, it is necessary to define a coding for the solution, so that it can be represented as a chromosome. Let $X = \{x_1, x_2, \ldots, x_n\}$ be the set of objects to add to a backpack, along with its respective sizes and costs. A representation for a chromosome in this case is given by a sequence of bits $b_1 b_2 \ldots b_n$ which indicate if x_i, for $i = 1, \ldots, n$, are included in the backpack (1) or not (0). Figure 9.2 shows a chromosome representation for this problem. The fitness function for this problem is exactly the same as the objective function, as seen on Eq. (7.6).

In each of the generations, the genetic algorithm makes a selection of the chromosomes that will reproduce. This selection occurs in a probabilistic fashion, but with a bias to selecting the best elements. To implement such a selection strategy, genetic algorithms use a technique called biased roulette (see Fig. 9.3). This method is based in the random selection of an element with probability directly proportional to the value of fitness for the corresponding chromosome. This can be explicitly defined as

$$p_i = \frac{f(x_i)}{\sum_{j=1}^{m} f(x_j)},$$

where m is the number of chromosomes. In this way, the objective is to simulate the mechanism used by natural selection, in which best fit individuals have greater probability of survival and reproduction.

The operators in genetic algorithms are used to perform modifications on selected chromosomes as a way of creating new elements of the population. They are also the method used to create a new generation of chromosomes from the previous generation. Several genetic operations can be defined, depending on the characteristics of the problem, but the classic operators used on genetic algorithms are the crossover and mutation operators.

In the classic crossover operator, two chromosomes are initially chosen through a selection algorithm. Then, it is necessary to define, in a randomized way, a position in the binary representation that will be called a crossover point. A new chromosome is created in the following way. The algorithm copies the first part of chromosome

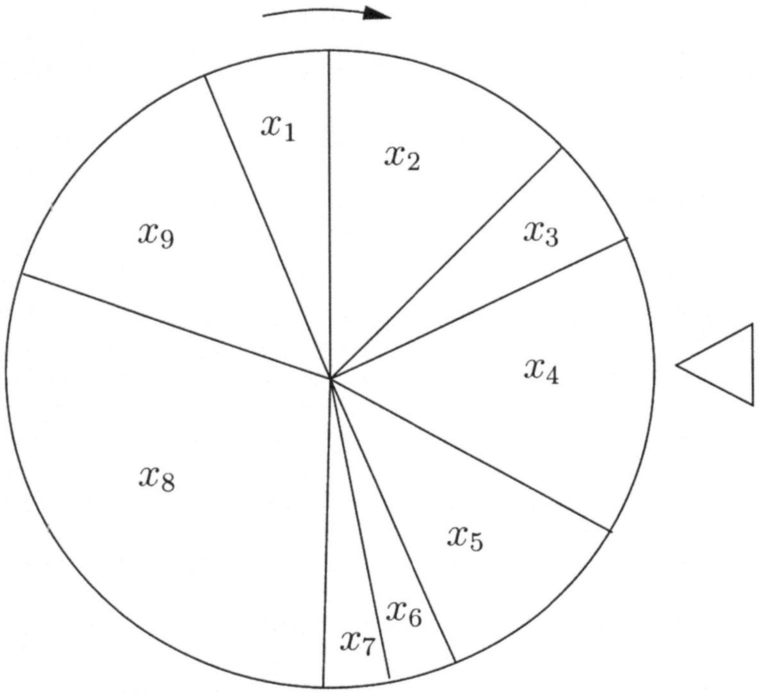

Fig. 9.3 The "biased roulette" selection strategy. Each chromosome has probability of being chosen that is proportional to its fitness value

1 up to the crossover point to the initial part of the new chromosome. Then, the last part of the second chromosome (starting from the crossover point) copies into the resulting chromosome. Because of the way it is constructed, the result will have a mix of the features present in the first and in the second chromosome. A result of the application of this operator can be seen in Fig. 9.4.

Starting from this basic crossover method, many variations can be adopted. For example, one can define operators that use more than one crossover points, selecting different parts from each of the generating chromosomes. One can also define crossover operators that take into consideration specific features of the problem. For example, in order to codify the traveling salesman problem as a chromosome, one can define a sequence of numbered cities that need to be visited. An efficient crossover operator needs to consider knowledge about this kind of representation in order to avoid repeated cities when different chromosomes are combined.

The second operator that is most frequently used in genetic algorithms is the mutation operator. This operator is used as a way of avoid local minima and to diversify the content of memory. Its main goal is to avoid stagnation of the current population of solution, which happens whenever the content of memory has the same or very similar solutions. The working of the mutation operator is simple,

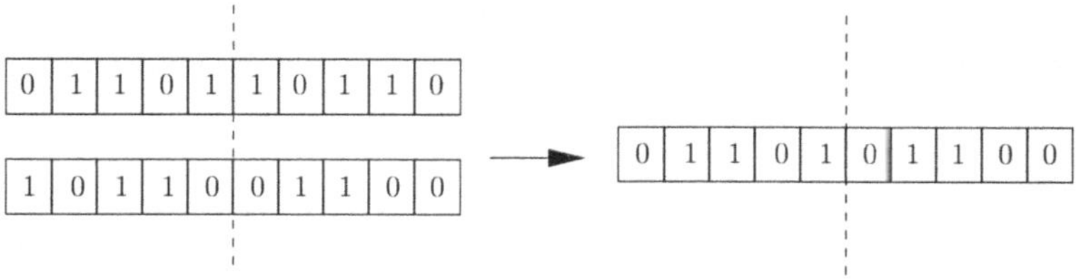

Fig. 9.4 Illustration of the crossover operator for the packing problem. In the left side, two chromosomes are presented. The dashed line indicates the crossover point. The right side displays the resulting chromosome

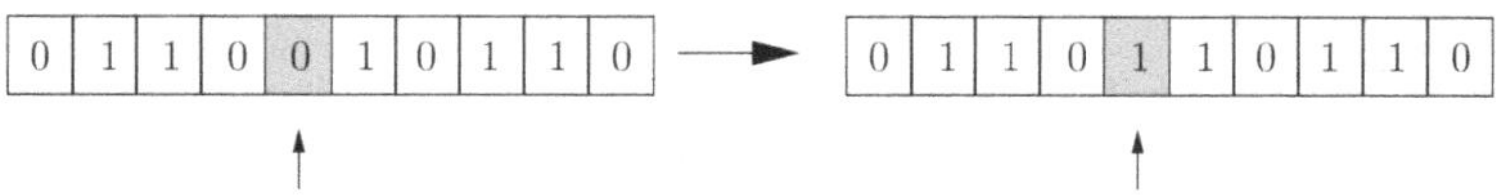

Fig. 9.5 Example of change performed by the mutation operator for a typical instance of the packing problem

Algorithm 13 Genetic algorithm metaheuristic

```
                                          ▷ Initialize population
Let population P ← ∅
Let i ← 0
while i <TAM_POP do
    Create new solution s
    Let P ← P ∪ {s}
    Let i ← i + 1
end while
Let i ← 0
while i <MAX_GEN do
    Selection procedure
    Crossover
    Mutation
    Update best solution s*
    Let i ← i + 1
end while
```

consisting of selecting one or more bits from the binary representation, and changing the bit to its inverse. An example can be seen in Fig. 9.5.

In the problem we used as a previous example, the packing problem, the execution of the mutation operator can randomly change the value of one of the elements from selected to unselected. The mutation operator should, in practice, be used infrequently, because it has the potential to make large changes to the existing solutions. Excessive use of mutation may defeat the purpose of the improvement process used by a genetic algorithm. In Algorithm 13, we show a pseudo-code description of a genetic algorithm implementation.

In practice, for many applications, these two operators (mutation and crossover) are not enough for the generation of good solutions. The main problem with these classic operators is that the modification performed by them frequently results in unfeasible solutions. This happens, because the solution is viewed by these operators as just a set of bits. In other words, these operators have no knowledge about the structure of the problem that is being encoded as a chromosome. A lot of the modern research in genetic algorithms is geared toward solving this difficulty. For an example of approach that may fix this issue, we have memetic algorithms [26], which try to integrate local search into genetic algorithms. Another approach for this problem is known as scatter search, which tries to improve genetic algorithms through the use of an improved form of nondeterministic search [14].

9.5 GRASP Metaheuristic

Greed Randomized Search or Greedy Randomized Adaptive Search Procedure (GRASP) is a heuristic algorithm used for solving combinatorial optimization problems. It combines elements of both greedy algorithms and randomness to find near-optimal solutions to complex problems. GRASP begins by generating a feasible solution using a greedy construction heuristic. The greedy algorithm constructs a partial solution by iteratively adding elements to it, considering a specific criterion, such as minimizing cost or maximizing value. This creates an initial solution, which may not be optimal but is a good starting point. The "randomized" aspect of GRASP comes into play after the greedy initialization. Instead of always selecting the best element to add to the solution, GRASP introduces an element of randomness by allowing some level of exploration. It selects from a limited set of candidate elements, potentially favoring those that are not the absolute best. This randomization aims to escape local optima and promote diversity in the solutions generated.

GRASP (Greedy Randomized Adaptive Search Procedure) is a metaheuristic proposed by Feo and Resende [7] and based on two main processes. First, there is a constructor for problem solution that has the main feature of generating good solutions in a randomized fashion [28]. Second, the metaheuristic relies on a local search heuristic [9], which acts as a method to improve solution quality.

GRASP has been successfully applied to a large number of combinatorial optimization problems, including problems such as scheduling [8], routing [11, 17, 23], facility location, quadratic assignment [24], knapsack [5], and maximum independent set, among other well-known optimization problems.

9.5.1 Main Features

The dynamic behavior of the GRASP metaheuristic is composed of a series of iterations. In each step, the constructor is executed, resulting in an initial solution for the problem. This solution is then passed on to the local search procedure. The

Algorithm 14 GRASP metaheuristic

```
1: Data input
2: while Optimality criteria not satisfied do
3:     Call greedy randomized constructor
4:     Call local search
5:     Update best solution
6: end while
```

final result of the GRASP algorithm is defined as the best solution found in each of the iterations. On Algorithm 14, a pseudo-code describing the main steps of the metaheuristic is provided.

After the data input on line 1 of the algorithm, GRASP stays on a loop that lasts, while the predefined stopping criterion is not reached. The selected stopping criterion can be any relevant measure, such as processing time, number of iterations, or even other application-specific measures of quality.

On line 3, the construction algorithm is executed. Then, on line 4, the local search algorithm is called. The local search algorithm has as its main purpose the improvement of the solution through a search of its local neighborhood.

Finally, on line, the solution is updated. In case a new solution with better quality has been found, the algorithm proceeds with storing the new solution as the incumbent optimum solution.

In the next sections, we describe in detail each of the major parts of the GRASP algorithm. We also provide additional information about the parameters used on the construction and improvement phase of this metaheuristic.

9.5.2 Intensification and Diversification on GRASP

During the process of designing new metaheuristics for combinatorial optimization, two strategies are often used. These strategies are responsible for the generation of solutions that are optimal or near-optimal. The two methods frequently used are:

1. *diversification*: This is a method used to avoid local minima, i.e., areas of the search space where the best point is not as good as the optimum solution.
2. *intensification*, which increases the possibility that a global optimum will be found, as long as the optimum value is part of the searched neighborhood.

Diversification is a strategy used in a metaheuristic to explore the solution space in a varied way. The goal of such a strategy is to scape local minima. The diversification of the search space normally implies on the acceptance of new solutions, even when they are worse than the solutions existing in the algorithm's memory. One can quickly provide examples of how this is done as other metaheuristics. Genetic algorithms, for example, provide diversification by the generation of new populations where the individuals are significantly different.

Another method used in GAs with this objective is the mutation operator. This operator creates individuals that have randomly assigned differences from the ones that are currently in memory. In a tabu search metaheuristic, the use of a list of tabu moves also has the objective of diversifying the search space. This also reduces the possibility of the algorithm moving to points of the search space that has already been explored. Finally, in the simulated annealing metaheuristic, the variation of temperature is the element that introduces diversification in new explored solutions. This happens because the higher the temperature parameter in the algorithm, the high the probability that a solution worse than the current one may be accepted by the algorithm.

Intensification is the strategy opposite to diversification. Its objective is to find the best solution that can be reached from a particular point in the solution space. In this way, intensification has the tendency of using greedy and myopic algorithms (i.e., algorithms that take into consideration only a small part of the search space). As examples of effective use of diversification, the metaheuristic tabu search will normally use movements that lead to the best neighbor solution in a given search space. Genetic algorithms employ the selection and crossover operators in a similar way, trying to intensify the features of good solutions stored in the population. Finally, in the simulated annealing metaheuristic, the reduction of temperature leads the algorithm into gradual convergence to a local optimum.

The GRASP metaheuristic was designed with the goal of providing fast convergence. Its central idea is to use intensification and diversification in the most direct way possible. The GRASP algorithm runs a loop that at each iteration calls a constructor, immediately followed by a local search algorithm in order to improve the generate solution. The constructor in GRASP works as diversification component, providing at each iteration a new starting point so that the local search can be restarted from different areas in the search space. Another feature associated with the GRASP constructor is the generation of solution which, in most of the cases, is of high quality. This contributes to reduce the computational effort of the local search algorithm and, consequently, improves the overall efficiency of the metaheuristic.

9.5.3 Construction Phase

GRASP uses a construction algorithm that builds initial solutions for the target problem. The constructor in GRASP is said to be greed, randomized, and adaptive. The constructor uses a greedy strategy that gives preference to high-quality solutions according to preestablished criteria. The algorithm is randomized, because the selection of elements that will become part of the solution is made using a stochastic method.

The stochastic selection is made in such a way that better solution has higher probability of being selected. The fact that selection is randomized makes it possible to incorporate some levels of diversification into the generated solution. Finally, the

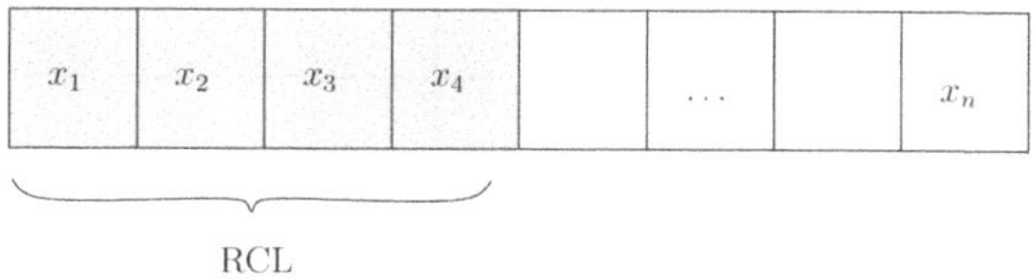

Fig. 9.6 List of available elements in a solution $(x_1, x_2, \ldots, x_n)$, sorted according to a fitness measure. Elements in the RCL $(x_1, \ldots, x_4)$ are the one defined by α parameter

Algorithm 15 Procedure for GRASP constructor

α parameter.
Let solution $S \leftarrow \emptyset$.
while Solution is incomplete **do**
    Sort candidate elements into a list (RCL)
    Select candidate x satisfying criterion determined by α parameter.
    Let $S \leftarrow S \cup \{x\}$
end while

constructor is adaptive, because the criteria for selection of elements during the execution of the algorithm are updated dynamically.

In most combinatorial optimization problems, a solution is formed by set of discrete elements. For example, in the case of the frequency assignment problem, a solution is constituted of a set of channels that are assigned to transmitters. Therefore, the elements that constitute a solution for the FAP are transmitters and channels.

The GRASP construction heuristic creates a solution incrementally. At each iteration of the construction algorithm, one of the best elements available at the moment is used. To avoid determinism, the constructor uses a list where available elements are stored. The elements considered for inclusion at each step are restricted to the ones that satisfy a greedy fitness parameter α. This list is called the Restricted Candidate List (RCL). Figure 9.6 shows a representation of the RCL with elements that satisfy the α parameter.

Algorithm 15 shows the pseudo-code for the construction in GRASP. After the initialization of RCL, the constructor selects in a random way one of the elements x from the list. The next steps consist of reordering the list of remaining elements according to the priority changes induced by the selection of x. This phase characterizes GRASP as an adaptive process, since the updates of RCL performed at each iteration. These steps are then performed until a complete solution is achieved.

9.5.4 *Local Search*

After the construction algorithm is executed, the next step in a GRASP metaheuristic is to use the generated solution as the starting point to a local search. This search has

Algorithm 16 Downhill algorithm

Initial solution s_i, number of attempts `MAX_TRY`.
Let $i \leftarrow 0$
Let $s \leftarrow s_i$
while $i <$ `MAX_TRY` **do**
    Let V be the neighborhood of s
    Select the element s' of V with lowest value of objective function.
    if $f(s') \leq f(s)$ **then**
        Let $s \leftarrow s'$
        Let $i \leftarrow i + 1$
    else
        Let $i \leftarrow 0$
    end if
end while

an objective to explore the neighborhood of the current solution, as a way of finding a local or global optimum, starting from that configuration.

There is a large degree of freedom for the definition of the local search methodology used in GRASP. It is necessary only that, given a valid solution for the problem, the algorithm is able to return an improved solution – or the same solution, in case that it is not possible to improve on it.

Several local search schemes for GRASP have been designed. Among them, one of the best known is the downhill algorithm, which is used in many other contexts. This algorithm is based on the use of local perturbations applied to a given solution. A local perturbation is any small change performed in a solution and that changes its objective function value. Starting from a specific solution, the set of solutions that can be found by the execution of exactly one local perturbation is called the neighborhood of the solution.

A description of the downhill algorithm in natural language is given in Algorithm 16. The first step in the algorithm is starting the solution and the repetition counter. Then, the algorithm enters a loop, where a number of tentative modifications will be performed in order to improve the objective value for the solution. This is performed by generating all elements in the neighborhood of the current solution and selecting the ones with lowest value of objective function cost.

Figure 9.7 shows an example of neighborhood for a solution s. If the new solution s' implies that $f(s') \leq f(s)$, then we update s to s' and continue to the next iteration of the loop. Otherwise, there was a failure in trying to improve the current solution — one can confirm this through the repetition count. When the number of failed improvement attempts surpass the value `MAX_TRY`, the loop ends. This algorithm has great advantages to the aspects of simplicity of implementation and efficiency.

9.6 Problems Solved by GRASP

Greedy Randomized Adaptive Search Procedure (GRASP) is a versatile optimization algorithm that has been applied to a wide range of combinatorial optimization

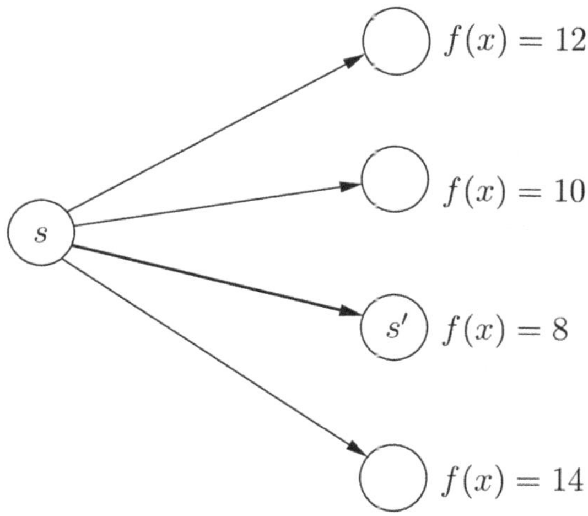

Fig. 9.7 Example of neighborhood for a solution s, during the execution of the downhill algorithm. Solution s' is a neighbor of s with lowest objective function value

problems. Here's a list of some problems that have been solved using GRASP, along with references to academic articles that demonstrate its application:

9.6.1 *Traveling Salesman Problem (TSP)*

The Traveling Salesman Problem (TSP) is one of the most well-known and extensively studied combinatorial optimization problems in the field of operations research and computer science [3]. The problem can be described as follows.

Given a set of cities and the distances between each pair of cities, the objective of the Traveling Salesman Problem is to find the shortest possible route that visits each city exactly once and returns to the starting city. This problem can be represented as a graph, where cities are nodes, and the distances between them are the edges.

The TSP is a classic example of a combinatorial optimization problem and is classified as an NP-hard problem. This means that as the number of cities (or nodes) increases, the number of possible routes (permutations) grows according to the factorial of the size of the problem, making it computationally infeasible to explore all possibilities for large instances.

9.6.2 *Job Scheduling*

The Job Scheduling Problem, also known as the Job Shop Scheduling Problem (JSSP), is a classic optimization problem in operations research and scheduling theory [27]. It deals with the allocation of limited resources to a set of jobs over time to optimize a specific objective, such as minimizing makespan, total completion time, or other performance metrics. The problem is particularly relevant in manufacturing, production, and project management contexts.

In the Job Scheduling Problem, you are given a set of jobs, each of which has a specific processing time and is required to be processed on a set of machines, each with its own processing speed. The goal is to determine the order in which jobs are processed on the machines to optimize a specific objective. Commonly used objectives in the JSSP include:

- Minimizing Makespan: The makespan is the total time required to complete all jobs, and the goal is to minimize this time.
- Minimizing Total Completion Time: The total completion time is the sum of completion times of all jobs, and the objective is to minimize this sum.
- Minimizing Total Tardiness: Tardiness represents the amount of time by which a job's completion time exceeds its due date. The objective is to minimize the total tardiness.
- Minimizing Maximum Lateness: The objective is to minimize the maximum lateness of any job.

9.6.3 Network Design

The Network Design Problem, also known as the Network Design Optimization Problem, is a class of combinatorial optimization problems that deal with designing an efficient and cost-effective network, such as a transportation or telecommunication network, to meet specific objectives. The problem involves determining the layout, capacity, and connectivity of the network's components to optimize performance metrics, including cost, reliability, and efficiency. Different network design problems arise in various domains, such as telecommunications, transportation, logistics, and infrastructure planning.

The specific characteristics and objectives of the Network Design Problem can vary, but the general problem can be defined as follows:

Given a set of nodes and a set of potential links or edges connecting these nodes, the Network Design Problem seeks to determine which links to establish, the capacity of each link, and the cost associated with these decisions. The objective is to optimize one or more criteria, such as:

- Cost Minimization: Minimizing the cost associated with establishing and maintaining the network, including costs of installation, maintenance, and operation.
- Reliability and Resilience: Ensuring that the network can tolerate failures or disruptions by adding redundancy or alternative paths while minimizing the cost.
- Efficiency: Maximizing network performance, such as throughput, capacity utilization, or minimizing congestion. Service Coverage: Ensuring that the network covers specific geographic areas or serves a certain population.

9.6.4 *Subproblems in Network Design*

The Network Design Problem also can be considered as a general problem, with several other subproblems. These subproblems find application in diverse areas of industry and services. A few specific examples include the following:

- Telecommunication Network Design: Designing telecommunications networks, including wired and wireless networks, data centers, and fiber-optic networks, to minimize latency, maximize data throughput, and minimize infrastructure and operational costs.
- Transportation Network Design: Designing transportation and distribution networks, such as road networks, public transit systems, or supply chain logistics, to optimize routes, minimize travel times, and reduce transportation costs.
- Internet Backbone Design: Optimizing the backbone structure of the internet to ensure efficient routing, minimize latency, and enhance network reliability.
- Wireless Sensor Network Deployment: Designing the deployment of wireless sensor networks for environmental monitoring, surveillance, or other applications while maximizing network lifetime, connectivity, and data coverage.

9.7 Summary

In this section, we considered several methods that combine search and artificial intelligence techniques to find solutions to hard combinatorial problems. Metaheuristics such as Tabu Search, Genetic Algorithms, and GRASP have been used to solve several optimization problems where the optimal solution is computationally expensive.

Applying metaheuristics to a combinatorial problem depends on finding efficient representations for the solution, and adapting an existing metaheuristic structure to the problem at hand. Many combinatorial problems have been already tackled by metaheuristics, showing the importance of this strategy to the solution of real-life optimization problems.

References

1. D. Abramson, Constructing school timetables using simulated annealing: sequential and parallel algorithms. Man. Sci. **37**, 98–113 (1991)
2. P. Carnevali, L. Coletti, S. Patarnello, Image processing by simulated annealing. IBM J. Res. Dev. **29**, 569–579 (1985)
3. P.F. Castro, F.G. Resende, C.C. Ribeiro, A grasp for the multiple traveling salesman problem. Eur. J. Oper. Res. (1996)
4. N.E. Collins, R.W. Eglese, B.L. Golden, Simulated annealing – an annotated bibliography. AJMMS **8**, 209–307 (1988)

5. A.T. Ernst, H. Jiang, M. Krishnamoorthy, Greedy adaptive search procedures for the knapsack problem. Comput. Oper. Res. (2001)
6. T.A. Feo, M.G.C. Resende, A probabilistic heuristic for a computationally difficult set covering problem. Oper. Res. Lett. **8**, 67–71 (1989)
7. T.A. Feo, M.G.C. Resende, Greedy randomized adaptive search procedures. J. Global Optim. **6**, 109–133 (1995)
8. T.A. Feo, K. Venkatraman, J.F. Bard, A GRASP for a difficult single machine scheduling problem. Comput. Oper. Res. **18**, 635–643 (1991)
9. M. A. Fischetti, J.S. Rosa, *Local Search in Combinatorial Optimization* (John Wiley & Sons, London, 1998)
10. M.R. Garey, D.S. Johnson, *Computers and Intractability – a Guide to the Theory of NP-Completeness* (W. H. Freeman and Company, 1979)
11. M. Gendreau, M.L. Rousseau, J.-Y. Potvin, Solving an aircraft maintenance routing problem with a grasp. Transp. Sci. (2000)
12. F. Glover, Tabu search - part I. ORSA J. Comput. **1**, 190–206 (1989)
13. F. Glover, Tabu search - part II. ORSA J. Comput. **2**, 4–32 (1989)
14. F. Glover, A template for scatter search and path relinking. *Artificial Evolution, Lecture Notes in Computer Science*, (1363) (1998), pp. 13–54
15. F. Glover, M. Laguna, *Tabu Search* (Kluwer Academic Publishers, Boston, 1997)
16. L. Goldstein, M.S. Waterman, Mappping dna by stochastic relaxation. Adv. Appl. Math. **8**, 194–207 (1987)
17. C.A. Hjorring, *The Vehicle Routing Problem and Local Search Metaheuristics*. Ph.D. Thesis, University of Auckland, Auckland, New Zealand, 1995
18. H. Holland, *Adaptation in Natural and Artificial Systems* (University of Michigan Press, Ann Arbor, 1975)
19. D.S. Johnson, C.R. Aragon, L.A. McGeoch, C. Schevon, Optimization by simulated annealing: and experimental evaluation; part I, graph partitioning. Oper. Res. (1992)
20. D.S. Johnson, C.R. Aragon, L.A. McGeoch, C. Schevon, Optimization by simulated annealing: and experimental evaluation; part II, graph coloring and number partitioning. Oper. Res. **39**(3), 378–406 (1991)
21. D.S. Johnson, C.R. Aragon, L.A. McGeoch, C. Schevon, Optimization by simulated annealing: and experimental evaluation; part III, the travelling salesman problem. Oper. Res. (1992).
22. S. Kirkpatrick, C.D. Gellat Jr., M.P.Vicchi, Optimization by simulated annealing. Science **220**, 671–680 (1983)
23. F.H. Lourenco, J.L. Roucairol, Applying grasp to the vehicle routing problem with split deliveries. Comput. Oper. Res. (1995)
24. R. Martí, P.M. Pardalos, Grasp with path-relinking for the quadratic assignment problem. INFORMS J. Comput. (2003)
25. K.K. Meyer, C.J. Nachtsheim, Constructing exact d-optimal experimental designs by simulated annealing. AJMMS **8**, 329–359 (1988)
26. P. Moscato, On evolution, search, optimization, genetic algorithms and martial arts. Technical report, California Institute of Technology, 1989
27. C.H. Papadimitriou, M. Yannakakis, Scheduling intersecting jobs on parallel machines. SIAM J. Comput. (1984)
28. L.S. Pitsoulis, P.M. Pardalos, Grasp: a greedy randomized adaptive search procedure. Comput. Oper. Res. (2002)
29. S. Porto, J.P. Kitajima, C.C. Ribeiro, Performance evaluation of a parallel tabu search task scheduling algorithm. Parallel Comput. **26**(1), 73–90 (2000). Special Issue on High-Performance Computing for Operational Research
30. I. Rechenberg, *Evolutionsstrategie: Optimierung technischer Systeme nach Prinzipien der biologischen Evolution* (Frommann-Holzboog Verlag, Stuttgart, 1973)
31. H.P. Schwefel, *Numerical Optimization of Computer Models* (Wiley, Chichester, 1981)
32. P.J.M van Laarhoren, E.H.L. Aarts, *Simulated Annealing: Theory and Applications*. (D. Reidel Publishing, Holanda, 1987)

Chapter 10
GRASP for Assignment Problem in Telecommunications

Carlos A. S. Oliveira

Abstract The Frequency Assignment Problem is a critical issue in telecommunications, where the goal is to assign frequencies to transmitters while minimizing interference across the network. Given the NP-hard nature of FAP, finding optimal solutions within polynomial time is infeasible, prompting the need for effective heuristic methods. Our focus is on adapting GRASP to address the specific challenges, including the management of local, co-channel, and adjacent channel interference. We describe the problem's formulation, the components of a typical FAP instance, and the objective function used to evaluate potential solutions. The chapter elaborates on the GRASP methodology, highlighting the use of restricted candidate lists (RCL) and iterative construction of solutions to enhance the quality and feasibility of frequency assignments. The effectiveness of the proposed GRASP-based approach is demonstrated through its ability to generate high-quality solutions that balance interference minimization with constraint satisfaction. By analyzing the results, we identify patterns and optimizations that contribute to improved performance in real-world applications. This work contributes to the broader field of combinatorial optimization in telecommunications engineering, offering insights into the application of metaheuristics for complex, large-scale problems.

Keywords Frequency assignment · Optimization · Greedy algorithms · Local search

10.1 Introduction

In this chapter, our focus will be on providing a comprehensive overview of the design and implementation of a greedy randomized adaptive search procedure (GRASP) tailored for addressing the Frequency Assignment Problem (FAP). As

C. A. S. Oliveira (✉)
Principal Inventive Scientist, Network Chief Technology Organization, AT&T Labs Inc., Bellevue, WA, USA
e-mail: carlos.oliveira@att.com

C. A. S. Oliveira, M. P. Pardalos (eds.), *Handbook of Artificial Intelligence and Data Sciences for Routing Problems*, Springer Optimization and Its Applications 219, https://doi.org/10.1007/978-3-031-78262-6_10

previously introduced in Chap. 8, the FAP aims to mitigate interference within the frequency spectrum by efficiently allocating channels to a network of base stations in cellular telephony systems. The central objective of the problem revolves around crafting an assignment function, denoted as $y : C \to \mathbb{N}$, which assigns transmitters to channels while ensuring that adjacent transmitters in the interference graph receive frequencies separated by a specified minimum distance, governed by the function $d : C \times C \to \mathbb{N}$.

The primary goal of the optimization process is to minimize overall interference, encompassing various types such as local interference, co-channel interference, and adjacent channel interference, as detailed in Sect. 6.3.3. It's important to note that the FAP lacks an efficient algorithm for finding optimal solutions within polynomial time, unless the complexity class P equals NP. Consequently, in this chapter, we delve into the application of GRASP as a metaheuristic approach for generating high-quality solutions to the FAP. The efficacy of GRASP lies in its ability to swiftly produce solutions of notable quality, a trait observed in similar problem domains. Subsequently, the subsequent chapter will dissect the outcomes derived from this GRASP-based methodology.

Throughout our discussion, we consider the details of GRASP and its adaptation to the features of the problem, considering mainly its performance and efficacy in solving real-world instances of the problem. By examining the results generated by this approach, we aim to discuss optimizations and potential avenues for further refinement in addressing the challenges posed by interference reduction in cellular telephony systems. This analysis is provided as a contribution to the growing body of research on heuristic methodologies for combinatorial optimization problems, particularly within the domain of telecommunications engineering.

10.2 Description of the Problem

An instance of the frequency assignment problem (FAP) is characterized by several key components. Firstly, it involves the determination of the number of transmitters ($n \in \mathbb{N}$) and the number of base stations ($m \in \mathbb{N}$) within the system. Additionally, crucial data includes a demand vector specifying the quantity of transmitters each base station possesses, denoted as m_i for $i \in 1, \ldots, m$. Moreover, the problem is further defined by two matrices: one representing the minimal channel separation between each pair of transmitters, denoted as $D_{i,j}$ where i and j represent base stations, and another matrix depicting the degree of co-channel interference ($C_{i,j}$) and adjacent-channel interference ($C_{i,j}$) between transmitter pairs. Both interference matrices contain real values ranging from 0 to 1, indicating the intensity of interference between base stations.

In essence, the FAP instance encapsulates the spatial configuration and communication dynamics within a cellular telephony system. It accounts for the distribution of transmitters among base stations, the spatial constraints imposed by minimal channel separation, and the varying degrees of interference between transmitter pairs, encompassing both co-channel and adjacent-channel interference. This comprehensive dataset forms the foundation upon which algorithms, such as the greedy randomized adaptive search procedure (GRASP), are deployed to optimize channel assignments, aiming to minimize interference and enhance the overall efficiency and reliability of the telecommunication network.

10.2.1 Instances and Objective Function

The initial phase of the GRASP algorithm starts with data entry In this phase, the data pertaining to the instance of the frequency assignment problem is read and validated. Information is stored in a instance description file, in a predefined format explained below.

An instance of the FAP is composed of the following data:

- The number of transmitters ($n \in \mathbb{N}$);
- The number of base stations ($m \in \mathbb{N}$);
- A *demand vector*, which says how many transmitters station base i has, for $i \in \{1, \ldots, m\}$ ($m_i \in \mathbb{N}$);
- A matrix $m \times m$ with values for minimal channel separation between each pair of transmitters ($D_{i,j} \in \mathbb{N}$);
- A matrix $m \times m$ with real values between 0 and 1, which represents the co-channel interference between each pair of transmitters ($C_{i,j} \in [0, 1]$);
- A matrix $m \times m$ with real values between 0 and 1, which represents the adjacent-channel interference between each pair of transmitters ($C_{i,j} \in [0, 1]$);

In general, it is hard to find algorithms that generate feasible solutions to the FAP, since even the problem of finding feasible solutions is known to be NP-Complete [3]. Therefore, it has been decided to use an objective function that takes into consideration the number of violations to restrictions in the given solution. The number of violated restrictions can be seen as a measure of feasibility for a solution.

The objective function $f(x)$, which the program seeks to minimize, is a composite expression integrating both the total interference within the system (8.1) and the count of constraint violations (8.2), each term weighted by a factor ρ. This formulation is devised to prioritize solutions that exhibit minimal interference while adhering closely to the specified constraints. By introducing the weight ρ, a higher emphasis is placed on penalizing solutions that deviate from the constraints, effectively incentivizing the algorithm to generate solutions that strike a balance between interference reduction and constraint satisfaction.

The function $f(x)$ is structured as a summation over various components: Firstly, a penalty term accounting for violations of minimum distance constraints, denoted by the condition $|y(v) - y(w)| < d(v, w)$, is multiplied by the weight ρ. Additionally, the function incorporates terms addressing co-channel interference and adjacent-channel interference, represented, respectively, by $c^{co}(v, w)$ and $c^{ad}(v, w)$. This comprehensive formulation captures the multidimensional nature of the optimization problem, enabling the algorithm to evaluate candidate solutions based on their ability to mitigate interference while minimizing constraint violations.

The objective function f which the program attempts to minimize if composed by a sum of the total interference in the system (8.1) and the number of violations in restrictions (8.2), multiplied by a weight ρ. In this way, a higher cost is assigned to a solution with more violated restrictions. Therefore, we have

$$f(x) = \rho \cdot \sum_{|y(v)-y(w)|<d(v,w)} 1 + \sum_{y(v)=y(w)} c^{co}(v, w) + \sum_{|y(v)-y(w)|=1} c^{ad}(v, w) \tag{10.1}$$

10.2.2 Constructor

The constructor used by GRASP is based on a modification of the T-Coloring algorithm , presented in Sect. 8.5.2.1. The main difference, however, between this and the T-Coloring algorithm is that instead of creating a complete solution, the algorithm only sorts the elements in the RCL in increasing order. The T-Coloring algorithm was used as the basis for the GRASP constructor, because it is a relatively simple method in comparison with other existing heuristics. Moreover, this algorithm is considered as one that offers high-quality solutions for the problem [3]. Finally, it was found in practice that the algorithm can be easily adapted to the alternative task of creating an RCL, an important step for the generation of solutions in a GRASP metaheuristic.

According to the methodology used by GRASP constructors, the complete solution is created in several iterations. In each iteration, the algorithm selects the next transmitter to which a frequency will be assigned. The criterion used for ranking transmitters is the number of available frequencies. The smaller the number of free frequencies available to be assigned to a specific transmitter, the higher its priority.

As explained in Sect. 9.5.3, the constructor avoids deterministic choices through the use of a restricted candidate list (RCL). In the proposed RCL, the transmitters are raked in increasing order of the number of available channel that can be assigned. Algorithm 1 shows a description of how the RCL is constructed.

Notice that this algorithm is very similar to Algorithm 9 for the T-Coloration heuristic, presented on Sect. 8.5.2.1.

Figure 10.1 shows an example that clarifies the ranking strategy used by the RCL. Antennas 1 and 2 need to receive one channel each. Antenna 1 has three neighbors with already assigned channels, while antenna 2 has only two neighbors in this situation. Consequently, antenna 1 has higher priority in the RCL, because it has less channels available for assignment (the three channels used by its neighbors are already in use). The idea of the algorithm is to guarantee that transmitters with less assignment options are processed first.

Algorithm 1 RCL creation algorithm

RCL is an empty list
Init **sat[i]** $\leftarrow 0, \forall i \in \{1, \ldots, |V|\}$.
Init **esp[i]** $\leftarrow 0, \forall i \in \{1, \ldots, |V|\}$.
Let $U \leftarrow V$.
while $U \neq \emptyset$ **do**
    Select $v_i \in U$ with largest combined value of **sat[i]** and **esp[i]**.
    Add v_i to the end of the RCL.
    Let $U \leftarrow U - \{v_i\}$.
    Update **sat[i]** and **esp[i]**, $\forall i \in \{1, \ldots, |V|\}$.
end while

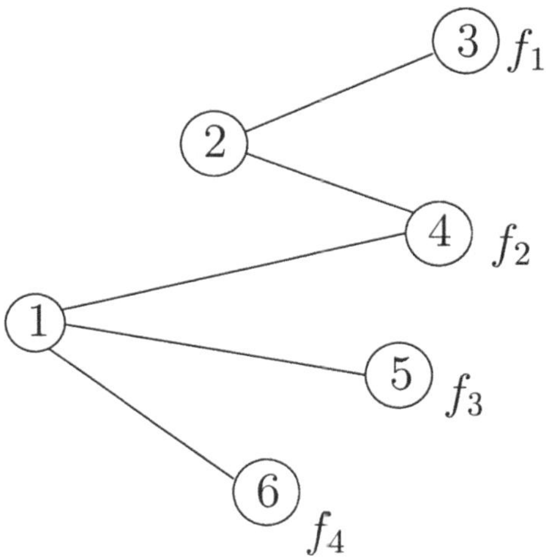

Fig. 10.1 Example for GRASP constructor

After the RCL is initialized, the next transmitter to receive an assignment is randomly selected from the first α elements. This way, even elements that are not among the best have a uniform probability of being selected. Finally, after a transmitter is selected, a channel is randomly assigned from the ones available at the moment.

10.2.3 *Local Search*

During the course of experiments, it has been shown that a different method for exploring the local structure of FAP instances would be more efficient. For example, instead of performing a perturbation in only one transmitter at a time, which frequently causes the violation of restrictions inside of a base station, it would be better to consider perturbations that encompass all transmitters located in a base station.

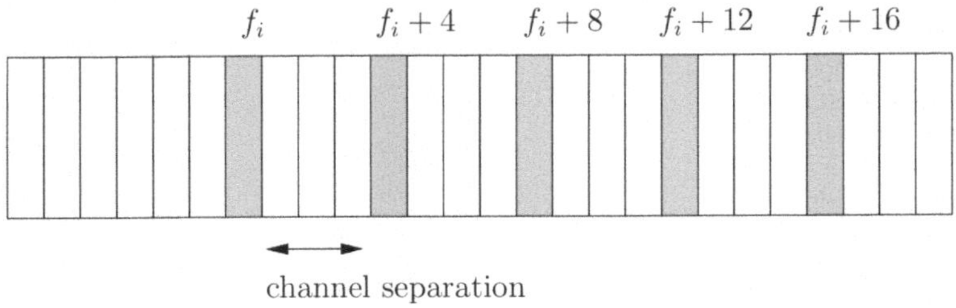

Fig. 10.2 Sample local search. The channels selected by the algorithm are marked as f_i, $f_i + 4$, $f_i + 8$, $f_i + 12$ and $f_i + 16$. The separation between channels is four units

Considering this, the algorithm used for the local search phase of the GRASP was created with the goal of exploring the structure of FAP cells. It is known that the transmitters in a mobile telephony system are grouped into base stations. Among them, local interference is measure as a constant value, which makes the minimum separation $d(i, j)$ a fixed quantity. Using this information, one can design a more efficient local search algorithm, assigning channel for the whole base station first, instead of doing this for each individual transmitter.

The algorithm is based in the downhill algorithm for local search, as described in Sect. 9.5.4. Two important features, necessary for the correct functioning of this algorithm, are the concepts of perturbation and neighborhood. A perturbation is a local change in the solution, which will frequently imply in a change of objective function value.

The neighborhood of a solution s is the set of solutions that can be found by a perturbation of s. Most of the work on the FAP that uses the concept of solution neighborhood assume that such a neighborhood is found by changing the channel assigned to a single transmitter. However, the type of perturbation used by the GRASP in this chapter assigns frequencies to a whole base station, instead of changing the channel for a single transmitter.

The main idea of the algorithm consists of, at each iteration, selecting a random base station. Next, one selects a band in the spectrum, also in a random way, which will provide the channels for the assignment to that base station. Figure 10.2 illustrates this process of channel selection. For example, in a base station with five transmitters, we will use five channels, separated by a fixed distance equal to the minimum separation needed to avoid local interference.

If this minimum separation is four and the initially selected channel is 40, then the next channels to be used are 44, 48, 52, and 56. When there are no additional channel available to assign to the whole base stations, new channels are added starting from channel 1. For example, suppose that in the above case the number of available channels was 50. Then, the resulting sequence would be $\{40, 44, 48, 2, 6\}$. In this way, the assignment of channels is performed to the whole-based station, and at the same time the feasibility of the choices is guaranteed, at least inside each base station.

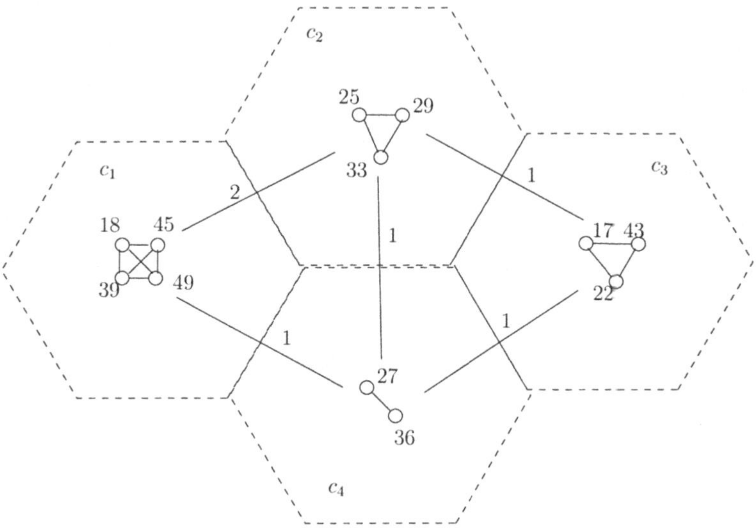

Fig. 10.3 Sample result for the local search algorithm Base station B was selected by the algorithm, and an In-block assignment is performed

An example of how the algorithm operates during local search is given on Fig. 10.3. In this figure, we see a simple instance of the frequency assignment problem, with only four base stations. To perform a perturbation of the solution, it is necessary to randomly select one of the base stations. In this example, station base B is the one selected. Right after that the algorithm selects a channel for assignment, which in the figure is channel 25. Suppose that the separation between channels is four units. Then, one can assign channels to the three transmitters that are part of this base station by simply taking consecutive channels that are separated by this minimum distance. The resulting channels in this example assignment are 25, 29, and 33.

10.3 Reactive GRASP

In a GRASP algorithm, the only parameter that needs to be determined in an arbitrary way is the value α, used during the construction phase. The value of α represents the number of elements in a restricted candidate list (RCL) which will be considered for the selection of a new element. Usually, finding the best value for α involves the execution of a set of experiments to calibrate the algorithm, with the objective of determining the most effective value for the considered problem. However, it may happen that α is dependent not only on the problem but also on characteristics of the instance we want to solve. In this case, the difficulty in determining α is even greater.

With the objective of providing more flexibility to the execution of GRASP and to facilitate the calibration of the algorithm, it would be better to determine α in an automatic way. In other words, given as input the instance we want to solve, it would be desirable to have an algorithm for optimal selection of α. With this objective, we adopted the method proposed by Prais and Ribeiro [12], known as *reactive GRASP*.

Reactive GRASP operates through the maintenance of a set of candidate values $\mathcal{A} = \{\alpha_1, \ldots, \alpha_n\}$ for the α parameter. Each item in $\mathcal{A}$ has an associated probability p_i. In each iteration of the reactive GRASP, a value α_i is selected from $\mathcal{A}$ with probability p_i and used in the constructor. Probabilities p_i are also updated after a certain number of iterations is such a way that these values may eventually converge toward optimal values of α.

The method in which probabilities p_i, used to select one of the α_i values, are updated may vary based on the application. A possible method has been proposed in [12], whereby a set $\overline{A} = \{a_1, \ldots, a_n\}$ is used, where each element a_i corresponds to the mean value of solutions found by the GRASP constructor throughout a number of executions, when the value α_i is used.

At the end of a predefined number of iterations, one can calculate a value q_i, for $i \in \{1, \ldots, n\}$, which represents the percentage from the best achieved when using α in the GRASP constructor. That is

$$q_i = \left(\frac{F(S^*)}{A_i} \right)^{\delta}.$$

Finally, probabilities q_i, for $i \in \{1, \ldots, n\}$, are updated as

$$p_i = q_i / \sum_{j=1}^{m} q_j.$$

The reactive GRASP strategy was implemented and applied to the FAP, as described above. We used the same steps described in [12] for calibrating the parameters.

10.4 GRASP and Path Relinking

Path relinking is a methodology of local search applied to metaheuristics. It was developed by Glover and Laguna [7] and initially applied to the tabu search metaheuristic. The main goal of path relinking is to promote the integration of the diversification and intensification strategies.

The methodology explores the path in the solution space traveled by the local search algorithm. Of great importance are the paths followed toward solutions with high quality. Such solutions are usually called *guiding solutions*. The use of this technique for improvement of GRASP solutions was proposed in [6]. Path relinking

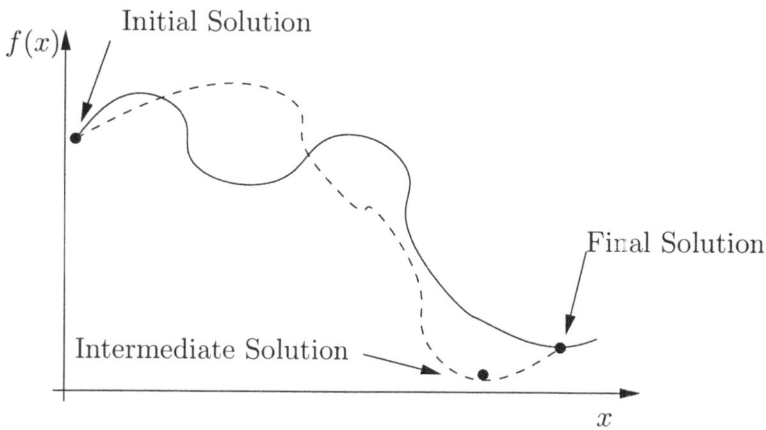

Fig. 10.4 Path relinking example. The full line represents the original path in the search space. The dashed line represents the new path following by path relinking when connecting the current solution to the guiding solution. The figure illustrates how the new path may pass through solutions that are even better than the guiding solution

is successfully used in a GRASP for the problem of minimizing crossings in the context of graph drawing [10].

Path relinking consists of using known high-quality solutions as guiding solutions. These solutions play the role of facilitators in the process of finding even better results for a given problem. Path relinking is able to find improved solutions by incorporating elements of existing high-quality solutions.

As apparent from the name of the heuristic, the main idea is to find a path through the solution space that connects the current solution to the high-quality solution that is being considered. During this process, it is possible that an even better solution can be uncovered as the search path gets closer to the existing high-quality solution.

Figure 10.4 shows an example of how this can happen. In the figure, the full line represents the original path followed through the search space in order to find the guiding solution. The dashed line represents the new path followed when one tries to reach the same guiding solution from the current solution. Between the initial and ending points, the search process can, at any point, reach a solution with objective function value that is better than the value for the guiding solution.

Another way of looking at the path relinking algorithm is considering it a method that make the original solution to look, at each step, more similar to the guiding solution. Even if they are very different at the beginning of the process, at the end, the current solution will have in the worst case an objective function value identical to the value of the guiding solution. However, there is also a possibility that a solution with even better fitness value be found before the current solution becomes identical to the guiding solution.

We decided to apply path relinking to our implementation of GRASP, as a way of improving the quality of generated solution. The path relinking methods allow

Algorithm 2 GRASP metaheuristic with path relinking

```
1: Input data
2: Let i ← 1
3: while Criterion not satisfied do
4:     Apply greedy randomized constructor
5:     Apply local search
6:     if i > PRSIZE then
7:         Run path relinking
8:     else
9:         add current solution to S_g
10:    end if
11:    Update best solution
12:    Let s̄ ← worst solution in S_g
13:    if f(s) < f(s̄) then
14:        Let S_g ← S_g s̄
15:        Let S_g ← S_g ∪ s
16:    end if
17:    Let i ← i + 1
18: end while
```

this to happen through the use of an additional step, applied only to one of the high solutions already found by previous iterations of GRASP. We describe the process in detail in the following paragraphs.

The execution of the path relinking method happens after the search phase of GRASP. At this moment, one can use the current solution as a starting point. Algorithm 2 shows the changes performed in the execution flow of a standard GRASP. One of the requirements for the execution of path relinking is the creation of a set of guiding solution, which are noted in the algorithm by S_g.

The number $n_g = |S_g|$ of solutions included in this set can be defined by a parameter, although a fixed number is usually adequate. During the first n_g GRASP iterations, no activity is performed by path relinking, since the initial content of S_g is still being created. As a result, one can see that the greater the size of S_g, the longer it takes before path relinking can give any contribution.

Updates to S_g will occur as high-quality solutions are found by GRASP in the subsequent iterations. As highlighted in line 13 of Algorithm 2, for each new solution found by GRASP, the algorithm checks if it has fitness value that is better than the worst value stored in S_g (we call the corresponding solution $\overline{s}$). If this is not true, then we insert $\overline{s}$ into S_g and remove the worst solution s.

10.4.1 Path Relinking Implementation

We still need to understand how path relinking is implemented for the frequency assignment problem. The necessary steps can be summarized as follows. Initially, an element of S_g is selected to become the guiding solution. This element is denoted

Algorithm 3 Path relinking

1: Initial solution s, with transmitters $\{t_1, t_2, \ldots, t_n\}$.
2: Randomly select $s' \in S_g$.
3: Let $\{t'_1, t'_2, \ldots, t'_n\}$ be the transmitters in s'.
4: Let $i \leftarrow 1$
5: **while** $i \leq |V|$ **do**
6:     Let $y(t_i) \leftarrow y(t'_i)$.
7:     Apply local search to s
8:     Let s^* be the best solution up to this moment.
9:     **if** $f(s) < f(s^*)$ **then**
10:         Let $s^* \leftarrow s$
11:     **end if**
12:     Let $i \leftarrow i + 1$
13: **end while**

by s', and the selection is made in a random way. Then, the algorithm starts a loop that executed for each transmitter in the solution.

The loop consists in changing the channel assigned to transmitter i in the candidate solution, so that it uses the same channel transmitter i uses in the guiding solution. By doing this, one allows that solution s becomes at each step more similar to s'. After the assignment is performed, the result can be improved through the application of a quick local search algorithm. This is what is displayed on line 6 of the algorithm.

Notice that there is the possibility that s may have worse fitness value after this channel exchange is performed. Therefore, local search is an important step that allows path relinking to find even better solutions. In the following steps, the algorithm compares the objective value found during path relinking with the best known value and makes updates if necessary. Finally, the algorithm goes back to the beginning of the loop, which is repeated until all transmitters have been considered.

The values of objective function found by path relinking can vary among the many executions of this loop. We know, however, that at each step, the solution s becomes more similar to s', and they will be the same at the end of the algorithm. Algorithm 3 shows a pseudo code for our implementation of path relinking.

10.5 Analyzing the Results

In this book, a number of heuristic and metaheuristic algorithms for the FAP have been discussed. Because of the impossibility of giving mathematical guarantees of performance for some of the heuristics discussed, computational experiments become an important part of such a work. In the following sections, we present computational results obtained by the GRASP discussed in the previous chapter.

Initially, we present in Sect. 10.9.1 the set of instances of the FAP used for comparisons in this work. Part of these instances were used as a validation method for the algorithm, as we discussed in Sect. 10.9.2. In Sect. 10.9.1, we describe how the instance generator for the frequency assignment problem was developed. Finally, we present results of tests performed with the implementation of GRASP in Sect. 10.10.

10.6 Instance Generation Problems

Generating instances of combinatorial optimization problems for research or testing purposes using Artificial Intelligence (AI) techniques is a fascinating area of study. It has been the subject of much research, as the increased need for algorithms to solve combinatorial problems escalate. In this section, we consider a few techniques to generate problem instances for combinatorial problems using AI-based algorithms.

The first step is to use AI-driven data generators to help create problem instances. For example, you can use AI models, such as generative adversarial networks (GANs) or variational autoencoders (VAEs), to generate synthetic problem data. The model can be trained on existing problem instances to generate new, realistic data that reflects the characteristics of the problem.

10.6.1 Generative Adversarial Networks

Generative adversarial networks (GANs) are a class of machine learning models designed for generating data, typically in the form of images, text, or other structured information. GANs were introduced by Ian Goodfellow and his colleagues in 2014 and have since become a groundbreaking development in the field of deep learning.

A GAN consists of two neural networks, each with a specific role:

- Generator (G): The generator network's primary function is to create data that is similar to, or indistinguishable from, real data. It takes random noise as input and generates samples that ideally mimic the data distribution of the training dataset. The generator essentially learns to produce counterfeit data.
- Discriminator (D): The discriminator network, on the other hand, is tasked with distinguishing between real data and data generated by the generator. It learns to classify input data as either real (from the training dataset) or fake (produced by the generator). Essentially, the discriminator acts as a binary classifier. The GAN framework operates as follows:

The generator creates synthetic data samples from random noise. The discriminator evaluates these generated samples and real data. The generator aims to produce better fake samples by minimizing the probability that the discriminator can correctly distinguish them from real data.

The discriminator tries to improve its ability to differentiate between real and fake data by maximizing this probability. These two networks engage in a training process that continues iteratively, each striving to outperform the other. As a result, the generator becomes more proficient at producing realistic data, while the discriminator becomes more discerning.

The training process is often described as a minimax game, because the two networks work against each other, leading to a balance where the generator produces increasingly realistic data. When the GAN training converges, the generator creates data that is almost indistinguishable from real data.

Applications of GANs are vast and include the following:

- Image Generation: GANs can generate high-quality, photorealistic images, making them useful in art, fashion, and design.
- Super-Resolution: GANs can enhance the resolution of images, making them valuable in medical imaging, photography, and video upscaling.
- Data Augmentation: GANs can create additional training data to improve machine learning models.
- Style Transfer: GANs can change the style of images, transforming photos into the style of famous painters, for instance.
- Text-to-Image Synthesis: GANs can generate images from textual descriptions.
- Video Generation: GANs can create realistic video sequences.
- Anomaly Detection: GANs can identify anomalies or outliers in data.
- Voice Generation: GANs can generate realistic-sounding human voice and music.

GANs have evolved significantly since their introduction, with many variations and improvements, including Conditional GANs, Wasserstein GANs, and Progressive GANs. They continue to be an active area of research and have led to remarkable advancements in the generation of creative content and the enhancement of various machine learning tasks.

10.6.2 Variational Autoencoders

Variational autoencoders (VAEs) are a type of generative model in machine learning and deep learning that are used for generative tasks, such as generating new data samples that are similar to a given dataset [9, 13]. VAEs combine elements from autoencoders and probabilistic modeling to create a model capable of generating data and performing efficient data compression and representation.

VAEs are structured like autoencoders, which consist of an encoder and a decoder. The encoder maps the input data into a lower-dimensional latent space, while the decoder maps data from the latent space back to the input space. This compression and decompression process is typically used for data reconstruction.

Unlike traditional autoencoders, VAEs introduce a probabilistic interpretation of the latent space. Rather than encoding data into a fixed point in the latent space,

VAEs map data into a probability distribution in the latent space. This allows for sampling different points within this distribution, effectively making VAEs generative.

VAEs introduce randomness into the model by sampling points from the distribution in the latent space. This randomness encourages the generation of new, similar data samples rather than just reproducing the input.

Another important characteristic is the objective function. VAEs are trained to optimize a specific objective function, which is a combination of two terms: A reconstruction loss, which encourages the generated data to be similar to the original data, making the autoencoder useful for data compression and denoising; and a regularization term, which minimizes the difference between the distribution in the latent space and a predefined prior distribution (usually a simple Gaussian distribution). This term encourages the latent space to be structured and continuous.

During training, the encoder maps input data to a mean and variance in the latent space. These parameters are used to define the probability distribution in the latent space, which can be sampled to generate new data points during the decoding process.

Finally, it is important to consider how to generate new data. To generate new data samples, one can sample from the latent space's distribution and use the decoder to map these samples back into the input space. This process allows VAEs to create new data samples that are similar to the data they were trained on.

Additional applications of variational autoencoders include:

- Image Generation: VAEs are used for generating new, realistic images. They have applications in image synthesis, style transfer, and data augmentation.
- Anomaly Detection: VAEs can detect anomalies by measuring the reconstruction error. Unusual data samples may result in higher reconstruction errors.
- Semi-supervised Learning: VAEs can be used in a semi-supervised setting, where labeled and unlabeled data are combined for learning. Data Compression: VAEs can be employed for data compression tasks.
- Representation Learning: VAEs learn useful representations of data in the latent space, which can be used for various downstream tasks.

10.7 Problem Representation and Feature Extraction

Choosing an appropriate representation is an important factor for the generation of problem instances. For many combinatorial optimization problems, this may involve defining the problem as a graph, matrix, or vector, depending on the nature of the problem.

If relevant, use AI techniques like dimensionality reduction [14, 16] or feature engineering to capture essential characteristics of the problem instances. This step can help you define the input features required for your problem.

One needs to define parameters that control the difficulty or complexity of the generated instances. AI can be used to adjust these parameters to create a range of instances with varying degrees of complexity.

10.8 Custom Generative Models

The development of custom AI models or algorithms can be tailored to the specific problem you want to generate instances for. These models should capture the inherent constraints and characteristics of the problem. For example, you might create a custom AI algorithm to generate instances of the traveling salesman problem or the knapsack problem.

One can, for example, use AI techniques to augment existing problem instances with variations. This can include perturbing existing data to create new instances or generating additional constraints for existing problems.

Another useful strategy is the use of randomization. This is a technique to introduce randomness into the instance generation process to create diverse problem instances. Randomization can be applied to the selection of parameters, the creation of constraints, or the introduction of noise.

Finally, validation and testing need to be done to ensure that the generated instances are valid and representative of the problem. This may involve checking that constraints are satisfied and that the instances adhere to known characteristics of the problem.

The use of generated instances to benchmark and evaluate optimization algorithms is warranted. One can then measure the performance of algorithms on these instances and use the results to assess algorithm efficiency and robustness.

The remaining step involves further tuning and refinement. The goal is to iterate and refine the instance generation process based on the results and feedback from algorithm evaluation. Adjust the parameters and constraints to create more diverse and challenging instances.

10.8.1 Open Access and Sharing

Finally, one should make the generated problem instances available to the research community. Sharing well-documented problem instances helps researchers compare algorithms and fosters collaboration in solving challenging problems.

By leveraging AI techniques, we can create a wide range of problem instances for combinatorial optimization problems. These instances can be used for research, algorithm development, testing, and benchmarking, ultimately advancing the field and facilitating the development of better optimization solutions.

10.9 Case Study: Frequency Assignment Instances

One of the issues that need to be addressed during the development of an optimization algorithm is the common lack of standard instances for a problem that can be used to measure the quality of proposed algorithms. For a fair assessment, it is necessary to use instances that are representative of the problem and at the same time that can serve as a referential for computational experiments. In this section, we describe the instances of the frequency assignment problem that were used in the tests presented in the remaining of the chapter.

To improve the reach of the performed tests, we used several instances with distinct characteristics. The types of instances considered for the execution of GRASP were the following:

- Initially, a set of randomized instances was used for our tests. This set was created using parameters that were known to occur in real-world instances of the frequency assignment problem. The random instances and the method used to generate them are described in Sect. 10.9.1.
- A second set of instances, where the optimum solutions are known beforehand, was used as a way of validating the results of our algorithm. These instances, called *reticulate instances*, will be described in Sect. 10.9.2.
- *Philadelphia* is an instance of the frequency assignment problem that has been discussed in many articles in the literature. It was used as one of the testing instances. The particular characteristics of this instance will be discussed in Sect. 10.9.3.

10.9.1 Random Instance Generation

One of the difficulties found by researchers when creating algorithms for combinatorial optimization problems is the lack of real-world instances, which can be used to compare different metaheuristic implementations. For the frequency assignment problem, the set of instances that is most well-known is the one made available by the CALMA project [1].

Despite this, the number of instance is small (11 at the last time we checked). A bigger problem for algorithm designers is that these instances are not for cellular telephony and therefore don't have some of the important characteristics that are unique for this model of telephony, such as the property that all antennas are grouped in only a few base stations.

Much of the work in the literature offer solutions for proprietary instances, usually provided by companies that own the cellular system installation [3, 5, 8]. One of the tasks performed during experimental phase was the generation of instances that had features similar to instances found in the real world.

In order to make this possible, information about instances such as provided in [3] were used. Instances described in that paper are private, owned by the German

Table 10.1 Parameters and values used by the FAP instance generator

Parameter	Minimum	Maximum
Number of nodes	267	4.240
Number of edges	20.164	529.000
Graph density (%)	6	[illegible]9[a]
Minimum node degree	0	[illegible]1
Mean node degree	66	249
Maximum node degree	174	561
Number of connected components	1	58
Maximum size of cliques in the graph	14	130
Number of edges (i, j) with $d(i, j) \neq 0$	1.053	29.524
Number of edges (i, j) with $d(i, j) = 1$	9	9.470
Number of edges (i, j) with $d(i, j) = 2$	1.046	17.934
Number of edges (i, j) with $d(i, j) = 3$	3	2.120
Number of edges with $c^{co}(i, j) > 0$	19.111	499.476
Number of edges with $c^{ad}(i, j) > 0$	233	103.290

[a] Only one instance had value greater than 57

company E-Plus. However, a set of relevant information about the instances were provided in the article, as illustrated in Table 10.1. Based on these values, an instance generator was created where the parameters for the output graph are in the same ranges reported in the article.

From the parameters shown in Table 10.1, only a few were considered for the generation of instances. The used parameters were graph density (ratio of total number of edges by maximum number of edges given by $n(n-1)/2$) and percentage of edges (i, j) with $d(i, j) = 1$, com $d(i, j) = 2$ e $d(i, j) = 3$. Finally, we also used information about percentage of edges with $c^{co}(i, j) > 0$ and $c^{ad}(i, j) > 0$.

With this data, we prepared an algorithm capable of generating instances with general characteristics similar to the ones presented in Table 10.1. The algorithm works by making random selections, but verifying the selections so that the instance won't cross the established parameters. Instances generated in this way are presented in Table 10.2. In this table, columns represent, respectively, the number of base stations and transmitters, density of the graph (ratio between total number of edges and maximum number of edges), and number of edges with $d(i, j) = 1$, $d(i, j) = 2$, $c^{co}(i, j) > 0$ and $c^{ad}(i, j) > 0$, respectively.

10.9.2 Generating Instances with Known Optimum

During the experimental phase, a set of validation instances were used to verify the correctness and convergence of the GRASP implementation. These instances have properties that allow us to known the exact value of the objective function for the

Table 10.2 Set of instances of the frequency assignment problem used in the experiments. Columns denominated e, n, and m represent, respectively, the number of base stations, the number of antennas, and the number of edges in the interference graph. The next columns represent the graph density (number of edges divided by maximum number of edges), and number of edges with $d(i, j) = 1$, $d(i, j) = 2$, $c^{co}(i, j) > 0$ e $c^{ad}(i, j) > 0$, respectively

	Parameters for instance generation				
e/n	dens(%)	$d(ij) = 1$	$d(ij) = 2$	$c^{co}(ij) > 0$	$c^{ad}(ij) > 0$
20/100	7	10	5	190	185
25/125	6	12	9	300	294
30/150	10	23	23	435	423
35/175	13	55	25	595	582
40/200	9	43	30	780	757
45/225	7	43	34	990	954
50/250	11	79	60	1225	1194
55/275	9	98	53	1485	1437
60/300	11	125	76	1770	1709
65/325	10	139	90	2080	2026
70/350	10	166	90	2415	2340
75/375	9	154	116	2775	2685
80/400	9	181	117	3160	3063
85/425	8	186	130	3570	3463
90/450	9	240	156	4005	3897
95/475	10	287	188	4465	4350
100/500	9	298	184	4950	4830
105/525	10	342	224	5460	5318
110/550	9	354	234	5995	5824
120/600	9	423	273	7140	6944

optimum solution. Creating such instances can be of great help in evaluating the convergence of the algorithms for the problem.

To evaluate this convergence, we created graph for which the optimal assignment is known. For example, for the first graph in Fig. 10.5, we known from graph theoretical properties that two colors are enough to assign different colors to each node in the graph. This is a common property to all graphs that have only cycles of even length.

As it is shown in Fig. 10.5a, all cycles have even length; therefore, it has this property. Similarly, it is known that three colors are enough to assign different colors to all nodes in the second graph of the figure, while four colors are sufficient for the third graph. Reticulated graphs, as shown in Fig. 10.5, will be called graphs of type 1, type 2, and type 3, respectively.

Using this information about the coloring number of a graph, one can create an optimal solution for the frequency assignment problem on the graphs displayed above, if we suppose that each node corresponds to a base station. For example, suppose that we have a system with 80 available channels.

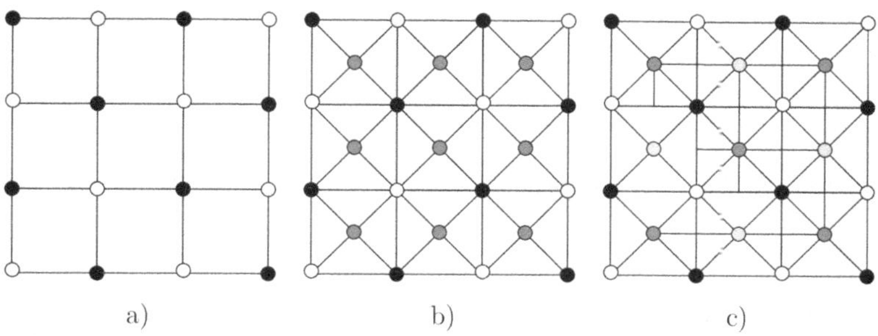

Fig. 10.5 Reticulated graphs. These graphs can be colored using only two, three, or four colors, respectively

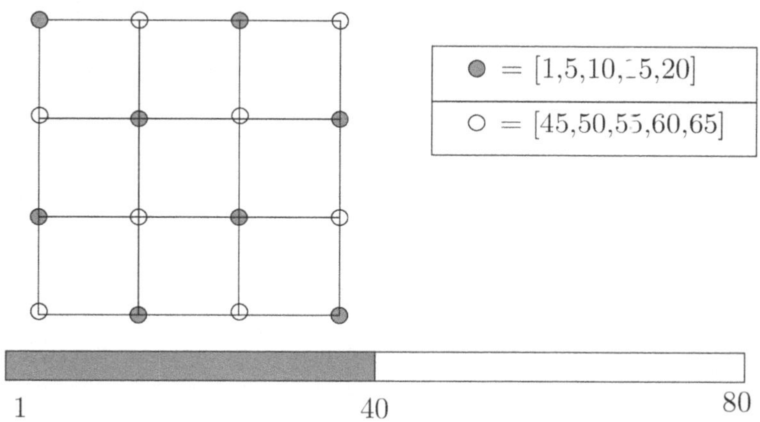

Fig. 10.6 Possible solution for the first reticulate graph shown in Fig. 10.5

One can generate an instance of the frequency assignment problem where the base stations are all graphs of type 1. As we need only two colors to create a proper coloration for such a graph, we can assume that the spectrum may be divided in two equal parts, as shown in Fig. 10.6. Each one of these parts will be referred to as a region of the spectrum.

One can also assume that the local, co-channel, and adjacent-channel interferences happen only along the edges of the interference graph. If we add a total of five transmitters to each base station, they can be freely assigned with colors from each one of the regions. A solution with zero interference for this simple graph can be found by using two color sets: one for the first regions of the spectrum $(1 \ldots 40)$ and another for the second region $(41 \ldots 80)$.

In the example shown on Fig. 10.6, the set of channels corresponding to color 1 is $\{1, 5, 10, 15, 20\}$ and the set corresponding to color 2 is $\{45, 50, 55, 60, 65\}$. In this example, there is a large amount of freedom for the choice of color groups, as long as the local interference restrictions are considered. However, as the available spectrum is reduced, the number of possibilities gets smaller. The same method can be used to generate instances with zero cost for graphs of type 2 and 3.

10.9.3 Philadelphia Instance

The instance of the frequency assignment problem known as Philadelphia refers to the configuration for a region around the city of the same name in the USA. This instance is composed by only 21 hexagonal cells, with the format as shown in Fig. 10.8. The demand vector is

$$m = (8, 25, 8, 8, 8, 15, 18, 52, 77, 28, 13, 15, 31, 15, 36, 57, 28, 8, 10, 13, 8).$$

The minimum distance restrictions between each pair of channels (i, j) are given by the following set of relations:

$$d(i, j) = \begin{cases} 5, \text{ if } i = j \\ 2, \text{ if } 0 < dist(v_i, v_j) \leq 1 \\ 1, \text{ if } 0 < dist(v_i, v_j) \leq 2\sqrt{3} \end{cases}$$

where $dist(x, y)$ is the Euclidian distance between x and y. These conditions result in the matrix presented in Fig. 10.7.

The importance of these instances resides in the fact that it was studied by many authors [2, 4, 11, 15, 17], who used it to test different approaches for the frequency

Fig. 10.7 Distance matrix for minimum distances on the "Philadelphia" instance

$$(d_{ij}) = \begin{pmatrix}
5&2&1&1&0&1&2&2&1&1&0&0&1&1&1&1&1&0&1&1&0\\
2&5&2&1&1&1&1&2&2&1&1&0&0&1&1&1&1&1&1&1&1\\
1&2&5&2&1&0&1&1&2&2&1&1&0&0&1&1&1&1&1&1&1\\
1&1&2&5&2&0&0&1&1&2&2&1&0&0&0&1&1&1&0&1&1\\
0&1&1&2&5&0&0&0&1&1&2&2&0&0&0&0&1&1&0&0&1\\
1&1&0&0&0&5&2&1&1&0&0&0&2&2&1&1&0&0&1&0&0\\
2&1&1&0&0&2&5&2&1&1&0&0&1&2&2&1&1&0&1&1&0\\
2&2&1&1&0&1&2&5&2&1&1&0&1&1&2&2&1&1&1&1&1\\
1&2&2&1&1&1&1&2&5&2&1&1&0&1&1&2&2&1&1&1&1\\
1&1&2&2&1&0&1&1&2&5&2&1&0&0&1&1&2&2&1&1&1\\
0&1&1&2&2&0&0&1&1&2&5&2&0&0&0&1&1&2&0&1&1\\
0&0&1&1&2&0&0&0&1&1&2&5&0&0&0&0&1&1&0&0&1\\
1&0&0&0&0&2&1&1&0&0&0&0&5&2&1&1&0&0&1&0&0\\
1&1&0&0&0&2&2&1&1&0&0&0&2&5&2&1&1&0&1&1&0\\
1&1&1&0&0&1&2&2&1&1&0&0&1&2&5&2&1&1&2&1&1\\
1&1&1&1&0&1&1&2&2&1&1&0&1&1&2&5&2&1&2&2&1\\
1&1&1&1&1&0&1&1&2&2&1&1&0&1&1&2&5&2&1&2&2\\
0&1&1&1&1&0&0&1&1&2&2&1&0&0&1&1&2&5&1&1&2\\
1&1&1&0&0&1&1&1&1&1&0&0&1&1&2&2&1&1&5&2&1\\
1&1&1&1&0&0&1&1&1&1&1&0&0&1&1&2&2&1&2&5&2\\
0&1&1&1&1&0&0&1&1&1&1&1&0&0&1&1&2&2&1&2&5
\end{pmatrix}$$

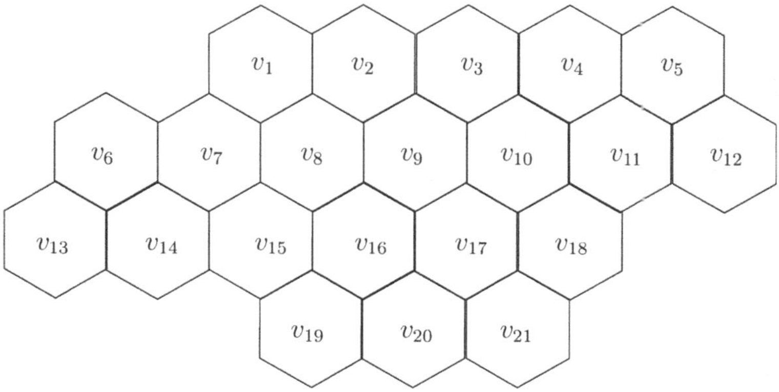

Fig. 10.8 "Philadelphia" instance

assignment problem. Therefore, it is interesting to use it as one of the test instances for the algorithms studied here. Moreover, this is one of the first real instances of the problem that were made publicly available for researchers (Fig. 10.8).

10.10 Computational Results

In this section, we present result of computational experiments performed with the discussed algorithms. The equipment used for testing included two workstations IBM RISC/6000 with 128 MB or memory each. The operating system used was AIX 4.2. Heuristics and metaheuristics discussed were coded in the C ANSI language.

The compiler used was gcc 2.95.1 with the optimization parameter `-O2`. All tests presented in this section were executed ten times, each execution with the duration of at most 1 hour.

10.10.1 Instances with Known Optimum

We tested the GRASP algorithm described in the previous chapter using instances generated according to the method of Sect. 10.9.2. The main goal of the initial experiment was verifying the time necessary to reach the known optimum solution. It should be noted that for the reticulated instances, the known optimum is equal to zero, i.e., these instances were designed, so they can be assigned without any interference. This information was used to validate the heuristics and verify the speed of convergence that could be achieved when applying the algorithm to real instances.

We generated sets of instances corresponding to the standard set of tests. The first set consists of reticulated graphs of type 1. The number of stations used corresponded to the graphs of a standard set. Similarly, a second set consists of reticulated graphs of type 2 and the third of reticulated graphs of type 3.

One can observe that the GRASP can find optimal solutions which is less time than other methods. Moreover, local search alone cannot find an optimal solution for the larger instances of the problem (starting with 400 transmitters), because it can be easily trapped into local optima. On the other hand, GRASP with path relinking was able to achieve the same results in less time, which shows that there was an improvement in performance due to the use of this strategy.

We also tested the same instances with a reactive GRASP. Despite showing results identical to the original GRASP, we noticed some degradation of running time under reactive GRASP. We attribute this to the additional time necessary for calibration of the internal parameter α. The advantages of reactive GRASP may be felt, however, when instances require longer execution time, which is not the case for this set of instances.

10.10.2 Random Instances and the Philadelphia Instance

In this section, we summarize the results achieved by GRASP for random instances and the Philadelphia instance. The number of channels used during the tests was 80 channels, which is a typical value. The tests performed with the greedy heuristic T-Coloring had two main goals:

1. A modified version of this heuristic is already part of GRASP, being used by the constructor to generate a list of restricted candidates.
2. The heuristic was previously used in [3] and has proved to be one of the best heuristics for the frequency assignment problem. With this experiment, the objective was verifying how much the implementation of GRASP could be better than the T-Coloring heuristic.

One can see that the results were highly favorable to GRASP, which was to be expected, since the T-Coloring heuristic is already part of the first phase of GRASP. Moreover, one can see that this heuristic achieves results that are much worse than the other methods. Probably this is due to the fact that the heuristic is not well adapted to the structure of the problem instances we are using, in comparison with the results produced by other heuristics.

It is useful to notice that in the objective function used, the number of violated restrictions has a higher weight than the value of interference. Therefore, the high values achieved by the T-Coloration heuristic indicate that this method is not efficient for the elimination of such violations.

Regarding the local search heuristic, we see that for small instances of the problem, the results achieved are comparable with the ones produced by GRASP. However, as the number of transmitters increases, also increases the difference between the results of these two algorithms. Being a greedy algorithm, local search has difficulty in avoiding local minima, which reduces its performance on larger instances.

Regarding the values found by the reactive GRASP, one can see that there were improvements in relation to the values obtained by GRASP in isolation.

Looking at the values found by the reactive GRASP, one can see that there was an improvement in relation to the results achieved by standard GRASP. Moreover, one can see in the last column of the table that the values for the α parameter are used for the iteration where the best solution was found. As reactive GRASP has more freedom to determine the value of α, it is able to perform a more complete exploration of the search space for this problem. Also, reactive GRASP permits the automatic identification of the best α parameter, so that it could be used later on a standard GRASP. One can see that the value 0.2, among the ones that were tested (the range goes from 0.1 to 1.0, with increments of 0.1), is the value that produces better results for the majority of the instances.

The path relinking strategy was added to GRASP with the goal of improving the quality of the generated solution. The results found by the application of reactive GRASP and path relinking show that the method resulted in an improvement for 14 out of 20 random instances that were tested. This translated the contribution brought by the additional search performed by path relinking. Despite this, there was a perceived increase in the time before the optimum solution was found. This was probably due to the additional effort needed to execute path relinking.

Figure 10.9 shows a graph summarizing the dynamic effect of GRASP with path relinking and reactive GRASP. This graph relates computational time to the value of objective function for random instances with 50, 60, 70, 80, 90, 100, 110, and 120 transmitters. It is noticeable that the main convergence of the algorithm happens quickly, around 30 seconds, even for larger instances of the frequency assignment problem. From then on, the improvements start to occur more slowly. However, notice that the best solutions are found after a relatively long time in comparison with the convergence time.

Figure 10.10 shows the evolution of objective function with respect to time for several methods. These results were obtained for a random instance with 600 transmitters. One can observe that local search is the method that displays slower convergence while at the same time finding solutions with much inferior quality in comparison to the other methods. Times for reactive GRASP and path relinking are also reported, presenting gradual quality improvements. The values found by the T-Coloring heuristic are not displayed in this graph, because they were worse than the values produced by all other methods, which would make it harder to visualize the rest of the graph.

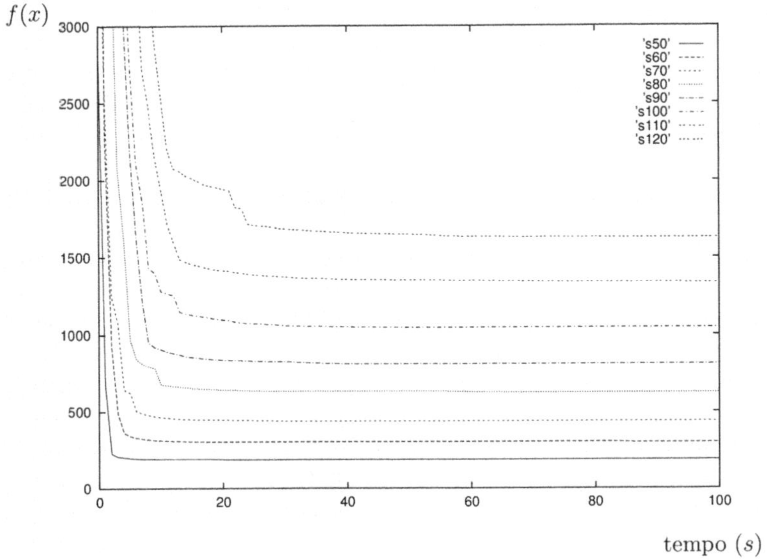

Fig. 10.9 Graph of the variation of best known objective function over time for reactive GRASP with path relinking. Tested instances had sizes 50, 60, 70, 80, 90, 100, 110, and 120 transmitters

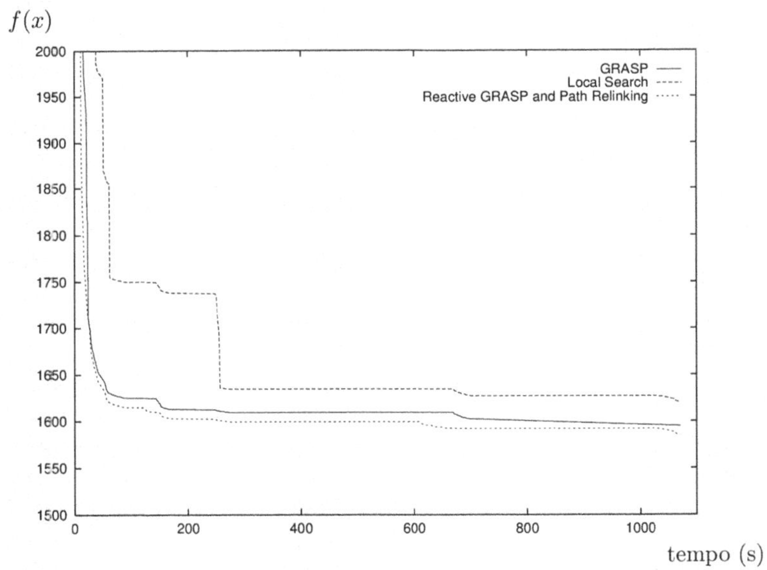

Fig. 10.10 Comparison of objective function values over time for an instance with 600 transmitters, using different methods

References

1. K. Aardal, C.A.J. Hurkens, J.K. Lenstra, S.R. Tiourine, An overview of algorithmic approaches to frequency assignment problems, CALMA project. Technical report, Dep. of Mathematics and Computer Science, Eindhoven University of Technology, Eindhoven, Holanda, 1995
2. L.G. Anderson, A simulation study of some dynamic channel assignment algorithms in a high capacity mobile telecommunications system. IEEE Trans. Commun. **COM-21**, 1294–1301 (1973)
3. R. Borndörfer, A. Eisenblätter, M. Grötschel, A. Martin, Frequency assignment in cellular phone networks. Technical report, Konrad-Zuze-Zentrum für Informationstechnik, Berlin, 1997
4. F. Box, A heuristic technique for assigning frequencies to mobile radio nets. IEEE Trans. Veh. Technol. **27**(2), 57–64 (1978)
5. A. Eisenblätter, A frequency assignment problem in cellular phone networks (extended abstract), in *Network Design, Connectivity and Facility Location*, volume 35 of *Dimacs Series in Discrete Mathematics and Theoretical Computer Science* (American Mathematical Society, 1997)
6. C. Fluerent, F. Glover, Improved constructive multistart strategies for the quadratic assignment problem using adaptive memory. Technical report, University of Colorado, 1997 Citado em [M. Laguna, R. Martí, GRASP and path relinking for 2-layer straight line crossing minimization. INFORMS J. Comput. **11**, 44–52 (1999)]
7. F. Glover, M. Laguna, *Tabu Search* (Kluwer Academic Publishers, Boston, 1997)
8. B. Jumard, T. Vovor, A column generation approach for the exact solution of channel assignment problems, in *Les Cahiers du GERAD* (1998)
9. D. P. Kingma, M. Welling, Auto-encoding variational Bayes, in *Proceedings of the 2nd International Conference on Learning Representations (ICLR)* (2014)
10. M. Laguna, R. Martí, GRASP and path relinking for 2-layer straight line crossing minimization. INFORMS J. Comput. **11**, 44–52 (1999)
11. R.A. Leese, Tiling methods for channel assignment in radio communication networks, in *3rd International Congress on Industrial and Applied Mathematics* (Hamburg, 1995)
12. M. Prais, C.C. Ribeiro, Reactive GRASP: an application to a matrix decomposition problem in TDMA traffic assignment. INFORMS J. Comput. **12**(3), 164–176 (2000)
13. D.J. Rezende, S. Mohamed, D. Wierstra, Stochastic backpropagation and approximate inference in deep generative models, in *Proceedings of the 31st International Conference on Machine Learning (ICML)* (2014)
14. S.T. Roweis, L.K. Saul, Nonlinear dimensionality reduction by locally linear embedding. Science **290**(5500), 2323–2326 (2000)
15. K.N. Sivarajan, R.J. McElice, J. W.Ketchum, Channel assignment in cellular radio, in *Proceedings od the 39th Conference IEEE Vehicular Technology Society* (1989), pp. 846–850
16. J.B. Tenenbaum, V. de Silva, J.C. Langford, A global geometric framework for nonlinear dimensionality reduction. Science **290**(5500), 2319–2323 (2000)
17. W. Wang, C.K. Rushford, An adaptive local-search algorithm for the channel assignment problem (CAP). IEEE Trans. Veh. Technol. **45**, 459–466 (1996)

Chapter 11
Waste Collection: Sectoring, Routing, and Scheduling for Challenging Services

Marcos Negreiros, Nelson Maculan, Claudio B. Cunha, Francisco Henrique Viana, and Flávio Luis Mello

Abstract This chapter shows old and new forms of waste collection that is being developed in different types of services performed and reported by the literature worldwide. Consequently, we indicate OR models and solutions that were reported and are here suggested. Services like domiciliary waste collection have been performed in different interesting methodological strategies in America, Italy, Brazil, and Argentina, as in other worldwide countries. Domiciliary selective waste collection is gaining recognition around the world as in Vietnam, Canada, and Brazil and are here discussed how they are being planned. Industrial, commercial and health waste collection are increasing annually, and they are introducing specialized routing and scheduling smart solutions with periodic components. At final, the

M. Negreiros (✉)
Universidade Estadual do Ceará, Laboratório de Computação Científica, Fortaleza, Brazil
e-mail: marcos.negreiros@uece.br

N. Maculan
Universidade Federal do Rio de Janeiro, Programa de Eng de Sistemas e Computaço (PESC-COPPE), Ilha do Fundão, Rio de Janeiro, Brazil
e-mail: maculan@cos.ufrj.br

C. B. Cunha
Universidade de São Paulo, Departamento de Eng de Transportes - Escola Politécnica, Ed Paula Souza (Prédio da Eng Civil) Cidade Universitária, São Paulo, Brazil

Graduate Program in Logistics Systems Engineering, University of São Paulo, São Paulo, SP, Brazil
e-mail: cbcunha@usp.br

F. H. Viana
Centro Federal de Educaço Tecnológica Celso Suckow da Fonseca CEFET/RJ – Campus de Nova Iguaçu Laboratório de Mobilidade Urbana Rio de Janeiro, Rio de Janeiro, Brazil
e-mail: francisco.viana@cefet-rj.br

F. L. Mello
Universidade Federal do Rio de Janeiro, Av. Brigadeiro Trompowiski s/n, Centro de Tecnologia, Bloco H, sala H-212b Cidade Universitária, Ilha do Fundão, Brazil
e-mail: fmello@poli.ufrj.br

C. A. S. Oliveira, M. P. Pardalos (eds.), *Handbook of Artificial Intelligence and Data Sciences for Routing Problems*, Springer Optimization and Its Applications 219, https://doi.org/10.1007/978-3-031-78262-6_11

substantial growth of the services in waste collection means a wide and complex management infrastructure that is here in discussion. The final impact of OR models in administrative and operational costs in the waste management structure is rethinking according to the institutional public decisions and summary.

Keywords Routing · Scheduling · Waste management · Optimization

11.1 Introduction

The municipal solid waste (MSW) management is an important task that local governments as well as private companies must consider protecting human health and the environment and to preserve natural resources.

To design an adequate MSW management plan, the first step consists of defining the waste generation and composition patterns of the town. As these patterns depend on several socioeconomic factors, it is advisable to organize them previously. Moreover, the waste generation and composition patterns may vary around the town and over the time. Generally, the data are not homogeneous around the city as the number of inhabitants is not constant nor it is the economic activity. The most important information about waste collection are street network and geographical accidents, population (spatial distribution, density, seasonal variation, economic level or HDI), economic activity (residential, commercial-type, industrial, mixed), and season [47].

The MSW is composed by many different types of services: domiciliary, selective, street sweeping (human and mechanical), container, commercial, industrial, health, fair, etc. These services used different approaches for daily planning and scheduling activities. Street-by-street routing for domiciliary and street sweeping, point-to-point routing for container, industrial, health, etc. [7, 10].

Previous literature revisions for waste collection sectoring and routing planning were developed by Eiselt et al. [43, 44] focusing on the postman-based algorithms and applications reported around the world; Ghiani et al. [49] focusing in the tactical and strategic issues related to most relevant Operations Research literature applied to solid waste management (SWM); Ghiani et al. [50] focusing in the waste management models for strategic (location and transportation with transshipment in single/multi-periods levels, others) and tactical levels (waste flow, districting and collection patterns); Han and Ponce-Cueto [60], focusing in the waste collection routing problems and their taxonomy; Sulemana et al. [117], focusing in routing literature for waste collection, and Hess et al. [63] focusing in the waste collection problems and their taxonomy.

In this study, we focus on the understanding of the principles of solving problems in different contexts of waste collection services (domiciliary, selective, commercial) and applications around the world, and along the reviews, we evaluate aspects of savings impacts, and finally we consider the design of the waste management system with sectoring and routing now available in the reported literature.

This chapter is organized as follows. In Sect. 11.2, the domiciliary waste collection is studied by means of its methodological aspects of solving different strategies to cover the daily production of garbage in distinct regions with their own features. In section 11.3 it is illustrated a round trip around some of major woldwide countries by continent explained how the domiciliary waste collection is planned and performed. In Sect. 11.4, we explain different aspects of performing selective waste collection in different countries. In Sect. 11.5, we present contexts in industrial, commercial, and health waste collection in different countries focusing in the differences. In Sect. 11.6, we show the features of a waste collection management system and some applications already being used and their main resources considering the proposed main features. In Sect. 11.7, we conclude this chapter with the main observations and consequent directions we had noticed during these study.

11.2 Domiciliary Waste Collection

In the waste collection structure, the planning of domiciliary waste collection (DWC) in major cities is made door-to-door, real-estate to real-estate, and street-segment by street-segment, and the organization of the service is very similar worldwide. The planning process is divided into strategic plans that guarantee the total city coverage with the minimum acceptable time by the local population, and using the minimum resources, the tactical plans consider the most effective form to deliver the contracted resources to cover by dividing the city into sectors with predicted behavior and minimum cost with vehicles, crews, and the routes to be performed, and finally the operational plans that consider the execution in the field to be manageable, traceable, and dynamically recoverable when interruptions occur [46, 47, 49].

The territory to be covered is previously given, the street network and their traversal constraints, streets that must be covered, local and special points that must be covered, places where the truck may not enter or must enter in certain direction, the estimate of garbage production per street/special collection places, the collection vehicle available (e.g., Load Packer), crew workload limits and average efficiency (ton/man-h), frequency and daily hour of service, and other important parameters may have to be previously obtained to prepare good plans [68]. To develop such plans, the specialized literature considers some of the following approaches, [43, 44, 49, 103]:

First Approach One way to develop good operational plans is to promote a giant route over the entire area and cluster the streets accordingly the vehicles' capacity and crew workload or form regions by clustering street or domiciles and perform the traveling paths in the formed regions. These methodologies differ mostly in the final solution, because the first way it can form regions that cross one another, or be much elongated, turning the areas difficult to manage the operational evolution of the service, and the routes are longer principally in major suburbs. The second one aggregate area making the routes be compact and restricted to an area with common behavior in waste production (in general good for suburbs with few one-way streets) but have the disadvantage of being difficult to accomplish in major downtown areas with many one-way street segments and nonhomogeneous load distribution because of the disparate size of the buildings and particular waste production.

Second Approach Regions and routes can be solved simultaneously, without steps. Although this method is closely related to the global mathematical optimization of any objective function associated (minimizing distance, total workload, fleet size and distance, and/or a mix of these objectives), in fact, this method is not conceived by any manager in waste collection. The decision process is in general according to the operational local difficulties, and many local constraints are solved by isolating and treating them as it may be adjusted with the reasoning process.

Third Approach Another way is to divide the process into two steps that are fundamental and mostly used by the literature. It devises the city into regions of the same frequency of visit (e.g., clustering or zoning) and applies a routing procedure to define the way of traversing these regions one by one that minimizes expenses with vehicles, crews, and equipment usage to transport the garbage in two phases: along the region (collecting) and from the region to destination (e.g., transfer station, landfill).

Fourth Approach Another way is promoted by municipalities for major Brazilian cities, where the regions are given and the concessionaires (e.g., contractors) are induced to define their costs at minimum to cover them all, defining or not the scheduling of service per region (alternate/diary, diurnal/night). In this case, the routes are designed to cover each street of the region with minimum cost, and if necessary, it can be broken any moment where the vehicle be overfilled. In this moment, the vehicle goes to the disposal site, discharge, and go back to the same point it was before continuing its route, then following this way until all the area is covered.

It is a common sense that the DWC is a problem related to the Chinese postman problem (CPP), [57], where it is simply given a strongly connected graph (with positive weighted in the links) representing the streets of a region and a vertex of this graph where a vehicle (e.g., load packer) departs and ends its journey to pass (collect and deadhead) through all the streets and returns to the starting point with minimum total distance of traveling (minimum deadheading).

The CPP is a well-studied problem in combinatorial optimization literature with many versions based on the graph structure: symmetric (where the graph is composed only by non-oriented edges), oriented (where the graph is composed only by oriented arcs), mixed (where the graph is composed by non-oriented and oriented links), windy (where the graph is composed only by non-oriented links but the way a link is traversed may return different traversal costs), and hierarchical (where the graph has links with preferences to be traversed than the others). Considering the general formulations reported by [37, 69] for the Mixed-CCP and [55] for the Windy-CPP [58], they can be stated as follows:

Sets:
V - Set of vertices.
A - Set of Arcs.
E - Set of Edges.
$L = \{E \cup A\}$ – Set of links.
$G(V, L) = G(V, E, A)$ – Mixed graph.
Parameters:
c_{ij} – Cost from vertex i to vertex j, $ij \in L$.
Variables:
x_{ij} – Number of times a link is traversed from vertex i to vertex j.

$$\textbf{(Mixed-CPP)} \min \quad \sum_{ij \in E} c_{ij}\, x_{ij} + \sum_{ji \in E} c_{ji}\, x_{ji} + \sum_{ij \in A} c_{ij}\, x_{ij} \tag{11.1}$$

s.t.

$$\sum_{ij \in L} x_{ij} = \sum_{ji \in L} x_{ji}, \; j \in V \tag{11.2}$$

$$x_{ij} \geq 1, \; ij \in A \tag{11.3}$$

$$x_{ij} + x_{ji} \geq 1, \; ij/ji \in E \tag{11.4}$$

$$x_{ij} \in \mathbb{Z}_{+}, \; ij \in L \tag{11.5}$$

Where (11.1) wants to minimize the total cost of the traversal of every link of the graph, constraint (11.2) is related to the flow conservation in the graph. Constraint (11.3) indicates every arc may be traversed at least once. Constraint (11.4) indicates every edge may be traversed at least once in one of its direction. Constraint (11.5) refers to the decision variables domain.

Example 11.1 In Fig. 11.1, we have a weighted mixed graph and its corresponding optimal solution multigraph obtained by running a B&B method to solve to the optimal its Mixed Chinese Postman instance [89]. This graph construction was done over a MapObjects GIS environment from ESRI™, where it is incorporated a graph editor, and its solution was obtained using the system XNES® RPP.

The windy CPP can be formulated as:

Sets:
V – Set of vertices.
E – Set of edges.
$G(V, E)$ – Windy graph.
Parameters:
c_{ij} – Cost from vertex i to vertex j, $\forall ij \in E$, consider $\exists\, c_{ij} \neq c_{ji}, \forall\, ij \in E$.

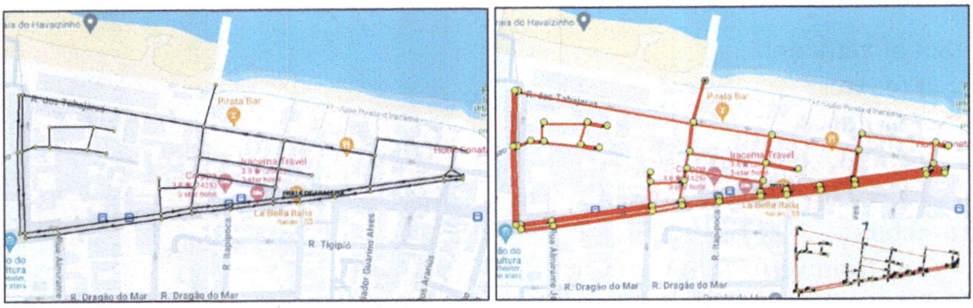

Fig. 11.1 A mixed weighed graph and the optimal MCCP multigraph from XNES®RPP, [89]

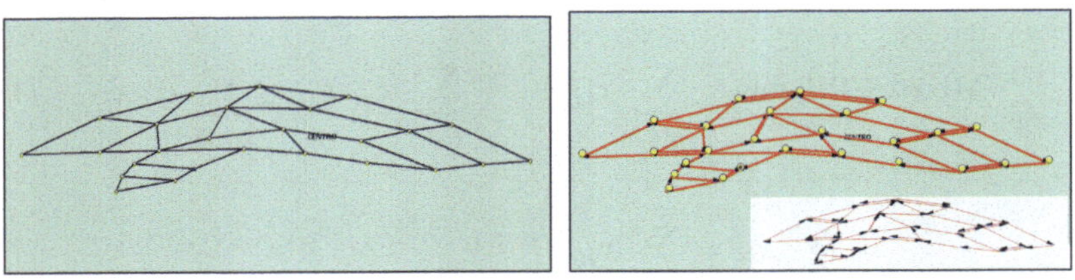

Fig. 11.2 A nonsymmetric weighed graph and the optimal windy CCP multigraph from XNES® RPP [89]

Variables:
x_{ij} – Number of times an edge ij is traversed from vertex i to vertex j.

$$\textbf{(Windy} - \textbf{CPP)}\ \min \quad \sum_{ij \in E} c_{ij}\, x_{ij} + \sum_{ji \in E} c_{ji}\, x_{ji} \tag{11.6}$$

such that:

$$\sum_{ij \in E} x_{ij} = \sum_{ji \in E} x_{ji}, \ j \in V \tag{11.7}$$

$$x_{ij} + x_{ji} \geq 1, \ (ij/ji) \in E \tag{11.8}$$

$$x_{ij} \in \mathbb{Z}_{+}, (ij/ji) \in E \tag{11.9}$$

Where (11.6) minimizes the total cost of the traversal of every edge of the graph at least once, constraint (11.7) is related to the flow conservation in the windy graph. Constraint (11.8) indicates an edge of the graph is traversed at least once in one of its directions. Constraint (11.9) refers to the problem's decision variables domain.

Example 11.2 In Fig. 11.2, we have a weighted windy graph and its corresponding optimal solution multigraph obtained by running a B&B method to solve optimally its Windy Chinese Postman instance [89]. The solution was obtained using the system XNES® RPP.

As can be seen, both models return the multigraph (a graph with repeated oriented links) that can be followed in such a way to define a Euler tour that pass only once by all links of the resulting Eulerian multigraph, [64], Fleury's algorithm – [76], Cycles – [54]. This means that it is possible to obtain a high combinatorial number of different identical cost solutions, Fig. 11.1. These solutions may have at least one of optimal traversal related to the effort of driving, or any other criteria. [16, 82, 90] proposed heuristic methods to find such tour maximizing vehicle in service, and minimizing turns as avoiding U-turns, turning the Euler tour execution acceptable by drivers. Although the literature went also to include in the movement prohibitions, the unacceptable traffic turns, where some turns can be considered forbidden and others more time-consuming or dangerous, especially for large waste collection vehicles. U-turns or certain right of left turns may be impossible to make due to narrow streets and/or are forbidden, because of this, authors consider modifying the original graph to deal with the problem and solving it as the asymmetrical TSP, [9, 36]. In [32], it shows that any of these transformations is necessary to solve arc routing problems based on CPP versions once it can be possible to maintain the quality of the solutions of the methods with little changes directed in the graphs without recurring to problem transformations as the previous step. [72] propose an arc-based shortest path method with forbidden turn and/or turn penalties that can easily solve arc routing by adjusting Dijkstra's method.

A step forward in CCP tours occurs when the literature turned to the general routing problem (GRP) and the rural postman problem (RPP)-related versions, [43, 100]. While the GRP is to perform a tour between required vertices and links, from a starting vertex (the garage) and back, in the RPP, the given strongly connected weighted graph has only a set of links (e.g., edges and/or arcs) that are required, and a tour with minimum cost departing and ending to the vertex v_0 correspond to the minimum rural postman tour. The first proposed model for the mixed RPP was introduced [36] and later confronted with another formulation in [36, 38]. The problem is proposed as follows:

Example 11.3 In Fig. 11.3, we have a weighted fully connected graph and its corresponding heuristic solution multigraph for a general routing instance. We used from SisRot® LIX to run this problem instance.

Given two subsets of vertices $S_1, S_2 \subseteq V$, $(S_1 : S_2)$ denotes the set of links with one endpoint in S_1 and other one in S_2, $A(S_1 : S_2) = \{ij \in A : i \in S_1, j \in S_2\}$, and $E(S_1 : S_2)$ denotes the set of edges with one endpoint in S_1 and the other one in S_2. Moreover, $\delta(S) = A(S : V\backslash S)$, $\delta^-(S) = A(V\backslash S : S)$, and $\delta(S) = E(S : V\backslash S)$, while $E(S)$ and $A(S)$ denote the sets of edges and arcs, respectively, with both endpoints in S. Furthermore, $\delta^*(S) = \delta(S) \cup \delta^+(S) \cup \delta^-(S)$. In order to denote the required links of a given set, subindex R will be used (e.g., $(S_1 : S_2)_R$) which represents the set of required links between S_1 and S_2).

A mixed graph $G = (V, E, A)$ is called *even* if all its vertices have *even degree* and *balanced* if, given any subset S of vertices, the difference between the number of arcs directed from S to $V\backslash S$, $|\delta^+(S)|$, and the number of arcs directed from $V\backslash S$

Fig. 11.3 A mixed weighed graph with required components (vertices and links), the optimal GRP route (multigraph) and a detail of the tour, from SisRot® LIX

to S, $|\delta^-(S)|$, is no greater than the number of edges joining S and $V \backslash S$, $\delta(S)$. The stronger proposed formulation for the MRPP is:

Sets:
V - Set of vertices.
E - Set of edges.
A - Set of arcs.
V_R – Set of required vertices.
E_R – Set of required edges.
A_R – Set of required arcs.
L – Set of links $L = E \cup A$.
L_R – Set of required links $L_R = E_R \cup A_R$.
$G(V, L, L_R)$ – RPP graph.
$(V^R) = \{V^1, \ldots, V^{|Q|}\}$ – Set of required connected components of G.
Parameters:
c_{ij} – Cost from vertex i to vertex j, $\forall ij \in L$.
Variables:
$x_{ij} : ij \in L$ representing the number of times a link $ij \in E \cup A$ is traversed.

$$\textbf{(MRPP)}\textit{Minimize} \quad \sum_{ij \in E} c_{ij}\, x_{ij} + \sum_{ji \in E} c_{ji}\, x_{ji} + \sum_{ij \in A} c_{ij}\, x_{ij} \quad (11.10)$$

such that:

$$x_{ij} + x_{ji} \geq 1, \; ij/ji \in E_R \quad (11.11)$$

$$x(\delta^+(S)) \geq 1, S \in \cup_{k \in Q} V_k, \; Q \subset \{1, \ldots, p\}, i \in V \quad (11.12)$$

$$x(\delta^+(S)) - x(\delta^-(S)) + \sum_{ij \in \delta(i)} (x_{ij} - x_{ji}) = 0. \quad (11.13)$$

$$x_{ij} \geq 0, x_{ij} \in \mathbb{Z}_+, ij \in A \backslash A_R \quad (11.14)$$

$$x_{ij} \geq 1 \text{ and integer}, ij \in A_R \quad (11.15)$$

$$x_{ij}, x_{ji} \geq 0 \text{ and integer}, ij \in E \quad (11.16)$$

Where (11.10) minimizes the *eulerian* and *balanced* multigraph, constraint (11.11) are the traversal inequalities. Constraint (11.12) imposes the resulting multigraph is connected. Constraint (11.13) force the multigraph be symmetric. Constraints (11.14, 11.15 and 11.16) define the space of the solution.

To deal with the extension of the RPP tour in downtown [9] and later [35] transform the directed or mixed rural postman problem to the asymmetrical traveling salesman problem (called Graphical Rural Postman Problem), because they also include traffic turn constraints to perform correct movements by drivers. The first math-heuristic proposed for the MRPP was developed by [33] and later [95]. Those are the ones available for the generalized arc routing context. Corberán et al [34] and [36] proposed exact cutting plane methods for the MRPP with success for moderate large instances.

Still, the CPP as RPP may not solve the real DWC problem in the majority of the route-first-cluster-second (RF-CS) cases, or contrary, although can be part of a solution for any case. The literature brings to us some different techniques in this direction: first cluster and routing or the contrary, and second simultaneously clustering and routing (applying the Capacitated Arc Routing Problem – carp, [53] [8]).

For the first approach, [17, 18] developed an initial approach that slightly generalize their proposed arc partitioning problem. In the given street network, each side of a street to be serviced is modeled as a separate arc. The length of an arc corresponds to the workload required to service it. If only one side of a street must be serviced, the graph is augmented by a parallel opposite arc with a length equal to the travel time along it. The resulting undirected multigraph is therefore Eulerian, and hence, the postman tour length is simply the sum of all edge lengths. In their heuristic, the authors first select a set of depots that serve as seeds for the districts. They assign simultaneously parallel pairs of streets to the depots. This way, each sector remains connected and Eulerian. After all edges have been assigned to a depot, exchange steps are applied to improve the balance of the sectors. A postman tour based on the non-augmented network is then computed in each sector to minimize deadheading.

Hanafi et al. [61] considered a sectoring problem for municipal solid waste collection and present a local search method to improve an already given sectoring plan. The underlying street network is represented by a mixed multigraph. Their problem contains two special features. First, the sectors are allowed to be disconnected, and second, because basic units may produce different amounts of waste, they consider daily varying collection and sectoring plans. This way, a basic unit can belong to different sectors on different days. However, within the same day, the assignment is unique. Because they allow disconnected sectors, the resulting routing problem is a rural postman problem. To simplify the problem, they approximate the routing distance based on the travel distances between the centers of the connected components of a district. In addition to the routing distance, they try to minimize the imbalance and the number of connected components of the sectors.

Mourao et al. [85] presented different heuristics for a waste collection problem. For their approaches, each side of a street that must be visited is modeled as a separate arc. A street with two sides that can be visited simultaneously is also modeled as two opposite arcs, only one of which is required. Furthermore, depending on the heuristic employed, arcs on the shortest path between odd degree vertices are inserted to obtain a Eulerian graph. The outcome of this preprocessing is a mixed multigraph. In their heuristic, the authors first select seeds for the sectors. At each iteration, they either add one single required arc or a small cycle to one sector. To achieve compact sectors, they use a closest assignment rule. The routing is either determined simultaneously with the construction of the districts or in a second phase after computing the sectors. The authors consider forbidden turns explicitly. As in the approach of [61], the set of required arcs of a sector may not be connected.

Simonetto and Borenstein [114] and later Li et al. [73, 74] proposed a model for the selective waste collection, considering: (a) the allocation of separate collection vehicles, as well as the determination of their routes, and (b) the determination of the daily amount of solid waste to be sent to each sorting unit to avoid waste of labor force and to reduce the amount of waste sent to the landfills.

For the second approach, the capacitated arc routing problem (CARP) was introduced by [53] and reports to a new way of solving the waste collection problem, by doing simultaneously routes and regions accordingly the vehicle's capacity and workload constraints. This problem returns solutions that can be reported to be optimal in cost of traversal, collection, and using the resources. The concept of required segments of a graph was here introduced, using the same concept proposed before by [100] when he defined the rural postman problem (RPP). The rapid development in this direction is notorious because of the problem generality interest for both arc and node routing literature [103].

Ulusoy [120] proposed an important approach to solve the problem considering also other constraints like crew workload and vehicle capacity for route first cluster second (RFCS). His method solves an optimal partition of a given sequence of visit, and [86] indicated that solving U-Turns and using [120] augmented graph returns solutions close to optimal in costs. This resolution is fully conditioned to the route-first step and may not solve appropriately the situation where a disposal site, different from the depot, is included in the decision process.

In the open CCP, it minimizes an open Euler tour cost that traverses all links of a graph starting in a departing vertex and ending in a different given vertex. Routes may be constructed for this case and levels of integration between two different movements (Garage $\rightarrow$ Region $\rightarrow$ Disposal Site and Disposal Site $\rightarrow$ Region $\rightarrow$ Disposal Site) in the same augmented graph may be considered for evaluation. As far as we know, no academic work considers this point for the CARP with intermediate facilities (CARPIF) or even for the sector-arc-routing (SARP) problems resolutions, although there is a direction that achieves very fast lower bounds for very large graphs (up to 9936 vertices, 16,877 edges, 10,213 arcs, 8373 required edges, 5084 and 2 intermediate facilities) for the CARP with intermediate facilities, [20]. Metaheuristic and Math-heuristic methods were developed with high-quality results for CARP instances commonly used by the literature lpr, mval, dea, gbd, eglese, and others, some corresponding to medium-sized cities, as can be seen in [23, 123, 127].

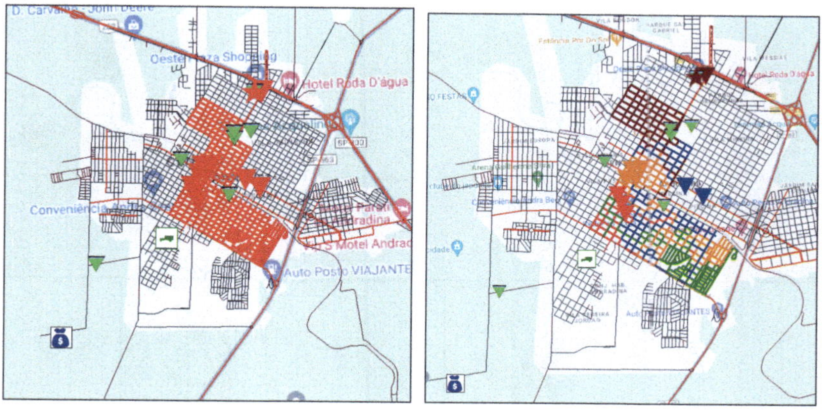

Fig. 11.4 A mixed weighed CARP graph (with required links), 7-ton vehicle 30t to collect and the best MCARP coverage, from SisRot ® LIX

Example 11.4 In Fig. 11.4, we have a weighted MCARP graph (for the city of Andradina/SP) and its corresponding solution multigraph (29.99-ton, vehicle 7-ton, balance between trips of 93.5%, 142.24 Km of total distance traveled in five trips) obtained by running a metaheuristic proposed by using Ulosoy's method starting with a MRPP heuristic method, [120]. In Fig. 11.5, it can be seen the routes performed and the dispersion of them once they do not take care of balance and compactness criteria. We used SisRot® LIX as the solver for this example.

For CARPIF, [126] present benchmark datasets for the mixed capacitated arc routing problem under time restrictions with intermediate facilities (MCARPTIF), which collection of instances is publicly available. Part of them refers to the municipalities of Centurion, South Africa, Actonville and Wattville, Benoni, and South Africa.

Duzgun et al. [42] proposed a transformation from the mixed capacitated general routing problem to the general vehicle routing problem that includes required vertices, edges, and/or arcs limited to vehicle capacity. Their solution method proved to deal with real-life data for compatible mixed networks. At that time, it was still a challenge to incorporate traffic flow conditions with turn penalties without changing the original graph. The original graph changes although in general is polynomial in major operations, and the cost of doing plans may be longer and constrained in the real context as cities grows. Limiting regions by frequency is a good way to reduce much the effort to make acceptable quality plans of this kind because of specific details related to local waste collection difficulties not related to mathematical or logical concerns.

It is important to note that the CARP solutions, even with intermediate facilities, may not warrant routes in contiguous and compact regions, meaning that they can be operationally difficult to control drivers and crews and/or it is difficult to manage municipality acceptability and waste collection execution. Although this happens,

Fig. 11.5 Visualization of the five routes from MCARP solution of the example 11.4, SisRot® LIX

there are works that report success in using CARP-like methodology as [105, 113] for the city of Coimbra/PT, [99] for Onitsha-Nigeria, [121, 122] for the city of Maringá-Brazil, [84] for the city of Oslo-Denmark, [65] for the district of Seixal/PT.

The most rich CARPIF mathematical formulations made at present were proposed by [110]. The authors proposed three different MILP formulations for: (1) The mixed capacitated arc routing problem with heterogeneous fleet and specialty link coverage (MCARP-HFSLC), where it reflects the situation in which a fleet, composed by three types of vehicles, leaves and returns to the same point every workday, and the points may have to be visited by a certain type of vehicle; (2) the Mixed Capacitated Arc Routing Problem with Limited Multi-Landfills (MCARP-LML); and (3) A MCARP-LML with two different vehicles. The third model was formulated as follows:

Sets:
V' – Set of vertices of intermediate facilities (IF) and the garage (D).
$V^* = V \cup V'$.
E – Set of edges.
A – Set of arcs.
V_R – Set of required vertices.
E_R – Set of required edges.
A_R – Set of required arcs.
L – Set of links $L = E \cup A$.
L_R – Set of required links $L_R = E_R \cup A_R$.

$L_{R1}, L_{R2}, L_R \backslash (L_{R1} \cup L_{R2})$, which, respectively, represent required arcs that must be served by vehicle type 1, required arcs that must be served by vehicle type 2, and required arcs that can be served by any vehicle.
$G(V^*, L, L_R)$ – Mixed CARP graph with intermediate facilities.
$(V^R) = \{V^1, \ldots, V^{|Q|}\}$ – Set of required connected components of G.
$IF \subset V$ – Set of intermediate facilities or landfills.
P – Set of vehicles of distinct kinds, $|P| = 2$ for this case.
τ – A trip between two vertices of V', where the first trip must start at the depot and the last must finish at the depot. When the vehicle leaves the garage and returns it must be empty. The vehicle is emptied $\tau - 1$ times at intermediate facilities.
τ_p – Number of trips vehicle type p.

Parameters:
c^1_{ij}, c^2_{ij} – Cost to serve a link ij with vehicle 1 or 2, respectively.
d^1_{ij}, d^2_{ij} – Cost of deadheading a link ij with vehicle 1 or 2, respectively.
q_{ij} – Demand/offer of the link ij, $q_{ij} = 0$ if the link ij is non-required.
v_1 – Garage.
λ^1_b, λ^2_b – Cost of emptying vehicle in the intermediate facility b with vehicle 1 and 2, respectively.
Q_p – Maximum capacity of vehicle type p.
P_p – Quantity of vehicles type p.
l_i – The number of allowed trips to the intermediate facility i.

Variables:

$$x^p_{ij} = \begin{cases} 1 \text{ if the link } ij \text{ is traversed in trip } t \\ 0 \text{ otherwise.} \end{cases}$$

ζ^t_{ij} is the quantity of waste collected during the trip t after crossing link ij.
y^t_{ij} - number of times link ij, on trip t, is traversed without collection.

(MHCARP-LIF)

$$\min \sum_{w=1}^{\tau_1} (\sum_{ij \in R} c^1_{ij} x^w_{ij} + \sum_{ij \in L} d^1_{ij} y^w_{ij} + \sum_{b \in IF: ib \in L} \lambda^1_b y^w_{ib}) + \sum_{k=\tau_1+1}^{\tau} (\sum_{ij \in R} c^2_{ij} x^k_{ij} + \sum_{ij \in L} d^2_{ij} y^k_{ij} + \sum_{b \in IF: ib \in L} \lambda^2_b y^k_{ib}) \tag{11.17}$$

s.t.

$$\sum_{t=1}^{\tau} (\sum_{ij \in L} y^t_{ij} + \sum_{ij \in R} x^t_{ij}) = \sum_{t=1}^{\tau} (\sum_{ji \in L} y^t_{ji} + \sum_{ji \in R} x^t_{ji}),\ i \in V^* \tag{11.18}$$

$$\sum_{ij \in L} y^t_{ij} + \sum_{ij \in R} x^t_{ij} = \sum_{ji \in L} y^t_{ji} + \sum_{ji \in R} x^t_{ji},\ i \in V,\ t = 1, \ldots, \tau \tag{11.19}$$

$$\sum_{t=1}^{\tau} x_{ij}^{t} = 1, ij \in A_R \tag{11.20}$$

$$\sum_{t=1}^{\tau} (x_{ij}^{t} + x_{ji}^{t}) = 1, ij \in E_R \tag{11.21}$$

$$\sum_{b \in V' : ib \in L} y_{ib}^{t} = 1, t = 1, \ldots, \tau \tag{11.22}$$

$$\sum_{b \in V' : bj \in L} y_{bj}^{t} = 1, t = 1, \ldots, \tau \tag{11.23}$$

$$\sum_{t=1}^{\tau} x_{ij}^{t} = 0, \ ij \in E_{\bar{R}} \tag{11.24}$$

$$x_{ij}^{\tau} = 0, \ ij \in E_R \tag{11.25}$$

$$\sum_{ib \in L} y_{ib}^{t} = \sum_{bj \in L} y_{bj}^{t+1}, t = 1, \ldots, \tau - 1, b \in IF \tag{11.26}$$

$$\sum_{t=1}^{\tau} \sum_{bj \in L} y_{bj}^{t+1} \leq l_b, \ b \in IF \tag{11.27}$$

$$\sum_{ib \in L} y_{ib}^{t} + \sum_{bj \in L} y_{bj}^{t} \leq 2, \ t = 1, \ldots, \tau, \ b \in V'. \tag{11.28}$$

$$\sum_{t=1}^{\tau} \sum_{b \in IF} \sum_{ib \in L} y_{bj}^{t} = \tau - 1. \tag{11.29}$$

$$\sum_{ij \in L} \zeta_{ij}^{t} + \sum_{ji \in L} \zeta_{ji}^{t} = \sum_{ij \in R} q_{ij} x_{ij}^{t}, i \in V, t = 1, \ldots, \tau \tag{11.30}$$

$$\sum_{bj \in L} \zeta_{bj}^{t} = 0, \ b \in IF, \ t = 1, \ldots, \tau \tag{11.31}$$

$$\zeta_{ij}^{w} \leq Q_1 \ (x_{ij}^{w} + y_{ij}^{w}), \ ij \in L, \ w = 1, \ldots, \tau_1 \tag{11.32}$$

$$\zeta_{ij}^{k} \leq Q_2 \ (x_{ij}^{k} + y_{ij}^{k}), \ ij \in L, \ k = \tau_1 + 1, \ldots, \tau \tag{11.33}$$

$$\sum_{w=1}^{\tau_1} x_{ij}^{w} = 0, \ ij \in A_{R2} \tag{11.34}$$

$$\sum_{k=\tau_1+1}^{\tau} x_{ij}^{k} = 0, \ ij \in A_{R1} \tag{11.35}$$

$$\sum_{w=1}^{\tau_1} (x_{ij}^w - x_{ji}^w) = 0, \; ij \in E_{R2}. \tag{11.36}$$

$$\sum_{i1 \in L} y_{i1}^{\tau_1} = 1, \; i \in V. \tag{11.37}$$

$$\sum_{1j \in L} y_{1j}^{\tau_1+1} = 1, \; j \in V. \tag{11.38}$$

$$\sum_{i1 \in L} y_{i1}^{\tau} = 1, \; i \in V. \tag{11.39}$$

$$x_{ij}^t \in \{0, 1\}, \; ij \in L, \; t = 1, \ldots, \tau \tag{11.40}$$

$$y_{ij}^t \geq 0 \text{ and integer}, \; ij \in L, \; t = 1, \ldots, \tau \tag{11.41}$$

$$\zeta_{ij}^t \geq 0, \; ij \in L, \; t = 1, \ldots, \tau \tag{11.42}$$

In the above formulation, the objective function (11.17) considers minimize the total sum of collecting, deadheading, and dumping. Constraints (11.18) and (11.19) impose the set of all trips must form a circuit. Constraints (11.20)–(11.21) guarantee that each required arc and edge must be served once. Constraints (11.22) and (11.23) guarantee that each trip enters and leaves at some disposal site just once, respectively. Constraint (11.24) imposes that a non-required edge is not served. In constraint (11.25), the garage is not a disposal site. In constraint (11.26) when a vehicle enters a disposal site for a certain trip, it must start the next trip at the same point. Constraints (11.27) and (11.28) guarantee the number of visits allowed for each disposal site is limited. Constraints (11.29) and (11.30) indicate that the amount of waste collected in a trip is updated after each service. Constraint (11.31) indicates that after leaving a disposal site, the vehicle must be empty. In constraints (11.32) and (11.33), the amount of waste collected cannot exceed the capacity of the vehicles T_1 and T_2, respectively. Constraints (11.34), (11.35), and (11.36) and vehicles T_1 and T_2 cannot serve certain arcs and edges. In constraints (11.37), (11.38), and (11.39), it is imposed to the vehicles to leave or return to garage in the respective trip. Constraints (11.40)–(11.42) are the domain of the decision variables.

Researchers that are developing solutions for the fourth approach are using the Sector-Arc Routing methodology, [40, 50, 85, 93, 110]. In this approach, the sectors and the routes are solved simultaneously or in two sequential steps, and any of the resolutions may consider that the routes (trips between garage → sector → destination → sector → ... → destination → garage) must be performed in the same region that have to be contiguous and compact. The contiguity measure is a criterion that indicates the streets of service may be mostly adjacent to each other, forming the minimum set of disconnected service components, while compact is related to the territory design, meaning that most part of the territory has no other inside; for contiguity and compactness indexes, refer to [24, 65, 110].

A general model proposed to sector routing for waste collection was introduced by [96] where they called: *Heterogeneous Arc Sector-Routing Problem* (HASRP). The model of the heterogeneous sector routing problem can be defined as a partitioning & assignment problem (PAP) as the following:

Sets:
S is the set of sectors;
L is the set of required street segments of an area;
V is the set of vehicles.

Parameters:
(t_v, Q_v) is a multi-set that defines the quantity per vehicle type and its maximum capacity;
W is the common crews workload;
c_s^v is the fixed/variable cost of assigning sector s using vehicle v;
q_{ij} is the garbage offer of the required street segment $ij \in L$;
w_{ij} is contribution in time of a required street segment $ij \in L$ to the workload;
$l_{ij}^{s,v}$ – Given status of a street segment $ij \in L$ that indicates it is used by the sector s with vehicle v, $l_{ij}^{s,v} \in \{0, 1\}$

Variables:

$$y_s^v = \begin{cases} 1, \text{ if sector } s \text{ is performed by vehicle } v \\ 0, \text{ otherwise} \end{cases}$$

$$(\mathbf{HSRP-PA}) \ \min \sum_{s \in S} \sum_{v \in V} c_s^v \, y_s^v \tag{11.43}$$

$$\text{s. t.:} \sum_{s \in S} \sum_{v \in V} l_{ij}^s y_s^v = 1, \forall ij \in L \tag{11.44}$$

$$\sum_{s \in S} y_s^v = 1, \ \forall v \in V \tag{11.45}$$

$$\sum_{v \in V} y_s^v = 1, \ \forall s \in S \tag{11.46}$$

$$y_s^v \in \{0, 1\}, \ \forall s \in S, \forall v \in V \tag{11.47}$$

The objective function (11.43) minimizes the total variable cost of assigning a route of sector s performed by vehicle v. The constraint (11.44) is a partition constraint that indicates that a street segment will be assigned to only one sector s and its related vehicle v. Constraints (11.45-11.46) are the assignment constraints, where constraint (11.45) indicates that just one vehicle is used per sector and constraint (11.46) indicates that just one sector is used per vehicle. The constraint (11.47) is the decision variables.

In this (*HSRP-PA*) formulation, it is important to notice that one sector may be described by different sets of routes composed by a feasible pattern of required street segments. These feasible patterns can be calculated such that:

$$\sum_{ij \in L} q_{ij}\, l_{ij}^{s,v} \leq Q_v^s, \ \forall s \in S, \ \forall v \in V \text{ and} \tag{11.48}$$

$$\sum_{ij \in L} w_{ij}\, l_{ij}^{s,v} \leq W, \ \forall s \in S. \tag{11.49}$$

To compose the columns of the partitioning part of the problem, it is necessary to use different mathematical programming methods for achieving feasible sectors to vehicles. To generate these sectors, one can handle it by using two auxiliary combinatorial optimization problems: heterogeneous capacitated centered clustering problem (*HCCCP*) quadratic version or a heterogeneous capacitated clustering problem (*HCCP*) a linear version, [91, 94], and for the selected links for each clustered sector, it is necessary to complete by applying the Urban Garbage Collection Problem, [88].

To design routes per sector, it was proposed by [88] the Urban Garbage Collection Problem following previous works based in the Canadian Postman Problem and Capacitated Arc Routing formulations, [53, 86, 88, 112]. This problem can be represented by using the following formulation:

Sets:
$G(N, L)$ is a graph composed by the street network vertices (N) and links (L) (oriented and non-oriented);
$L = (E \cup A)$ mixed links (arcs and edges);
$L_R = (E_R \cup A_R)$ is the set of required street segments of the graph G, composed by the required edges and arcs for service;
$L_R \subseteq L$ L_R is a proper subset of L;
V is the set of vehicles;
$A(i) = \{j \in N | (i, j) \in L\}$;
$B(i) = \{j \in N | (j, i) \in L\}$;
$|L|$ is the cardinality of the set L.

Parameters:
$\{1\}$ is the garage vertex;
$\{n\}$ is the disposal site vertex;
Q_{max} is the maximum load of a trip;
W is the maximum workload;
d_{ij}^t is cost of servicing link $ij \in L$ in trip t;
$stime_{ij}^t$ is time of servicing link $ij \in L$ in trip t;
$dtime_{ij}^t$ is the deadheading time of passing through a link $ij \in L$ in trip t;
ST^t is the stop time spent in the disposal site on trip t;
q_{ij} is the garbage household offer of the street segment $ij \in L$;

nt is the maximum number of trips to be covered by the vehicle in the sector.

Variables:
$f_{ij}^t \geq 0$ is the flow in the arc $< i, j > \in G$ in trip t.
$x_{ij}^t \geq 0$ is the number of times the vehicle pass through the link $ij \in L$ in trip t;

$$l_{ij}^t = \begin{cases} 1, \text{ if the street segment } ij \in L \text{ is serviced in trip } t \\ 0, \text{ otherwise} \end{cases}$$

$$\textbf{(UGCP)} \ \min \ z \tag{11.50}$$

s.t.

$$z \geq \sum_{ij \in S} \sum_{t \in T} (x_{ij}^t \ dtime_{ij} + l_{ij}^t \ stime_{ij}) + \sum_{i \in B(n)} \sum_{t \in T} ST^t \ x_{i,n}^t \tag{11.51}$$

$$\sum_{j \in A(i)} x_{ij}^t = \sum_{j \in B(i)} x_{ji}^t, \tag{11.52}$$

$$\sum_{j \in A(1)} x_{1j}^t = 1, \ t \in T \tag{11.53}$$

$$\sum_{j \in B(n)} x_{j1}^t = \bar{l}^t, \ t \in T \tag{11.54}$$

$$\bar{l}^t \geq l_{ij}^t, \forall ij \in A_R, \ \ t \in T \tag{11.55}$$

$$\bar{l}^t \geq l_{ij}^t + l_{ji}^t, \ ij \in E_R, \ t \in T \tag{11.56}$$

$$\sum_{j \in A(n)} x_{nj}^t - \sum_{j \in B(n)} x_{jn}^t = 0, \tag{11.57}$$

$$\sum_{t \in T} (l_{ij}^t + l_{ji}^t) = 1, \ ij \in L_R \tag{11.58}$$

$$l_{ij}^t \leq x_{ij}^s, \ ij \in L_R, \ \ t \in T \tag{11.59}$$

$$\sum_{ij \in L_R} q_{ij} \ l_{ij}^t \leq Q_{max}, \ ij \in L_R, \ \ t \in T \tag{11.60}$$

$$\sum_{j \in A(i)} f_{ij}^t - \sum_{j \in B(i)} f_{ji}^t = \sum_{j \in A(i)} l_{ij}^t, \ i \in N - \{n\} \tag{11.61}$$

$$\sum_{j\in A(n)} f_{nj}^t - \sum_{j\in B(n)} f_{jn}^t = - \sum_{ij\in L_R} l_{ij}^t, \ t \in T \tag{11.62}$$

$$f_{ij}^t \leq |L_R| x_{ij}^t, \forall ij \in L, \ \ t \in T, \tag{11.63}$$

$$f_{ij}^t \geq 0, \ ij \in L, \ t \in T \tag{11.64}$$

$$x_{ij}^t \in \aleph, l_{ij}^t \in \{0, 1\}, \ ij \in L, \ \ t \in T \tag{11.65}$$

The objective function (11.50) minimizes the time spent to cover the required set of street segments and deadhead. The constraints (11.51–11.57) warrant that the crossing by each vertex is even less the ones related to the garage and disposal site. The constraint (11.58) states that the service will be done in only one direction, by only one trip of the required set of links. The constraint (11.59) indicates that the vehicle will cross a required link servicing or not it at trip t. The constraint (11.60) is related to the loading of the vehicle from i to j limited to its maximum capacity. The constraints (11.61–11.63) are Golden and Wong [53] constraints to avoid disconnected infeasible sub-routes which were adapted to this UGCP model. Constraints (11.63) refer to the range of the problem decision variables.

Although the workload is defined to be at maximum W for covering a circuit, it is not used in the UGCP model, because one wishes to minimize the total time spent in performing all required links of the sector. It is assumed that the sector can be done in time to complete the workload; in any case if it is not possible, the necessary overtime is cleared minimized.

By using this latter approach, the concessionaires and municipalities may live peacefully, what it is in most Brazilian, Argentina, Czech Republic, and Portugal examples reported in next section. It happens because it facilitates operation and execution as reporting and following the execution by the population.

Example 11.5 In Fig. 11.6, we have a mixed SCARP weighted graph instance (for the city of Andradina/SP). In Fig. 11.7, it is shown the corresponding solution multi-graph (29.99-ton, vehicle 7-ton, 3 sectors, 6 trips, balance between routes of 77.50% and 154.96 Km of total distance traveled) obtained by running a metaheuristic proposed by Negreiros et al. [96].

It can be seen in Fig. 11.8, for the same instance, that the solution of the trips is more compact, a little less balanced and more contiguous than previous MCARP and SCARP solutions, but with a higher total distance than CARP. In Fig. 11.8 it is shown SARP corresponding solution with 6 trips, 87.99% ballarce and 151.79Km. The solution of the instance was also obtained with SisRot® LIX (Fig. 11.9).

Comparing solutions obtained in Examples 11.4 and 11.5, it is possible to notice that the effect of compactness and contiguously areas is obtained by specialized constrained clustering methods applied in different stages of the process of sectoring and routing. The effect of more compact trips helps in the final cost of the trips for this instance. Table 11.1 contains the resume of relevant aspects of the quality of both solutions for SCARP and SARP approaches.

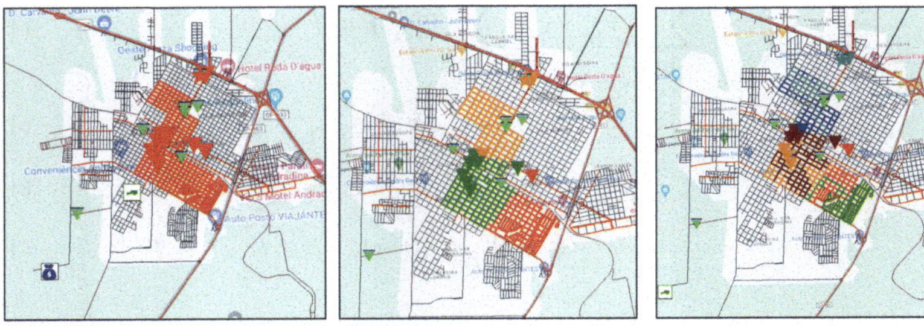

Fig. 11.6 A mixed weighed SCARP graph (with required links), 7-ton vehicle 29.99-ton to collect and the achieved SCARP coverage for example 11.5, from SisRot® LIX

Fig. 11.7 Visualization of the 6 trips from MSARP solution of the example 11.5, from SisRot® LIX

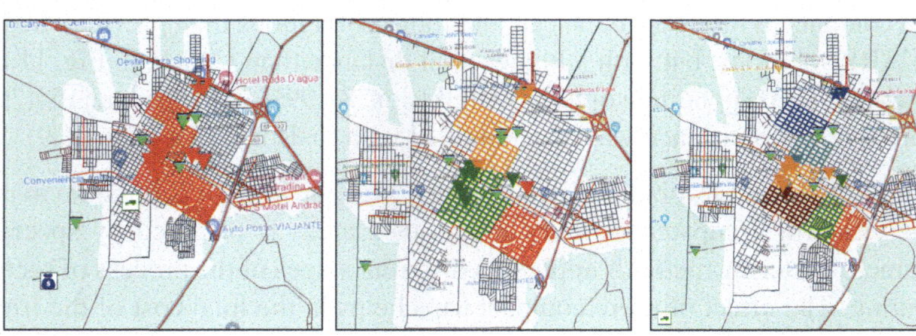

Fig. 11.8 A mixed weighed SARP graph (with required links), 7-ton vehicle 30t to collect and a new more compact coverage, from SisRot® LIX

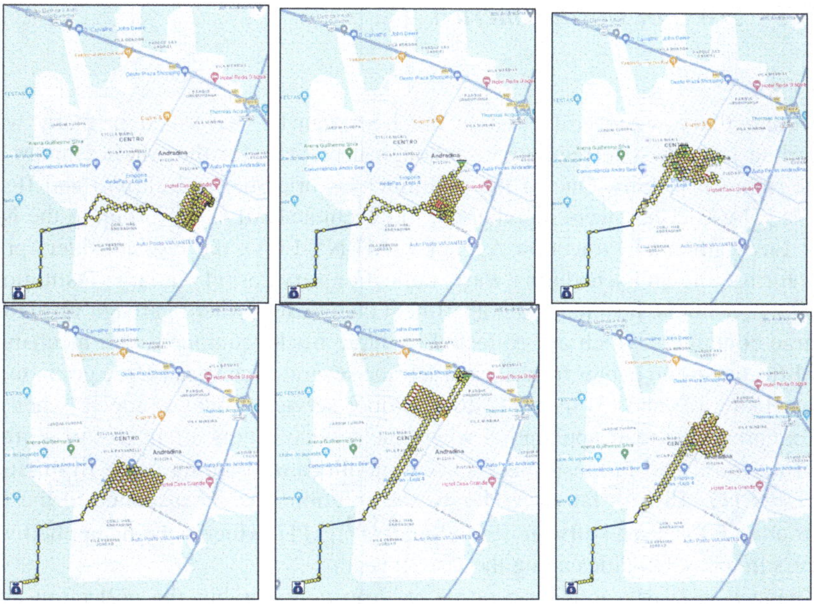

Fig. 11.9 Visualization of a more compact SARP approach to the example, from SisRot® LIX

Table 11.1 SCARP and SARP's different methods behavior and solutions, considering the effect of compactness, balance, and contiguity of trips in sectors

	Sector					
	01		02		03	
Method	Km	Bal %	Km	Bal %	Km	Bal %
SCARP	50.39	82.27	50.70	81 40	54.18	85.22
SARP	51.35	95.89	49.61	92 42	50.83	99.40

11.3 Domiciliary Waste Collection Around the World

In the following subsections, we present a specific selected overview of reported results and applications in domiciliary waste collection. Our tour starts in North America where the domiciliary waste collection is mostly dominated by the bin door-to-door system, and then we go to Central America and South America where the collection system varies between bins and continuous collection, from country to country or even in the same country; Europe and Africa, mostly dominated by bin door-to-door collection, but also appearing concentrated container and mobile system; Asia, where it is mostly performed in concentrated containers, and in some parts handcraft collection; and finally Africa, where it is mostly used door-to-door bin collection.

11.3.1 DWC in North America

The USA

The domiciliary waste collection in the USA can be performed in many ways according to each state regulation, although it prevails for the whole country the waste removal in bins. One of the preliminary works due to Beltrami and Bodin (1974) [7] while they were members of the Technical Advisory Group to the New York Environmental Protection Agency (TAG-NY-EPA), the city considers problems for routing and scheduling waste collection services: (1) in large institutional sites (e.g., schools, hospitals, and apartment complexes) where garbage was stored in large containers which are collected by large trucks equipped with mechanical fork lifts, hoist-compactor trucks, each service several such sites before going to the disposal site; (2) and (3) in DWC door-to-door service. Changes appear when the authors presented their application where the service occurs along arcs of a mixed street network, and propositions for location-allocation-routing came from Bodin and Levy [17, 18]. As far as we know, the solution was encapsulated in the first arc-routing GIS-based software RouteSmart^{TM} [111], which is being applied with success in many US cities along the last 30 years.

Changes in waste collection occur to drop considerably the collection costs imposed by crews' labor regulations, late organic and food waste state, and municipal regulations. The domiciliary waste turned in many US cities to source-separated organic collection in bins. The most recent redesigns of the conventional load packers were focused on three components: no conventional packing blades, welded leak-proof seams, and robust lifting. The vehicles look like a dump truck but can haul 10 tons of organic waste before it is full. With no push blade and waterproof welded seams, the waste is totally contained. This is an issue as it cannot have leaking when it is serving the food producer, grocery, and restaurant customers. Traditional manual collection requires at least two workers, the driver and the bin collector. Semiautomated waste collection is like traditional manual collection as the collector manually aligns the cart, which is then robotically emptied into the truck. Alternatively, fully automated collection requires only a single worker to drive and control and automated arm to collect recycling carts.

For these types of collection, the location of the carts is given in the street maps, and the amount of waste is forecasted for each. The set of locations is identified by a close-enough arc routing problem that selects the required arcs passing through streets close to the carts to minimize the routes. This problem called Profitable Close Enough ARP (PCEARP) was first investigated by [59] and [15] which formulates a MILP model that is solved to optimally for several synthetic instances; later, [109] proposed MS-VNS, GRASP-IT, and GRASP-BP metaheuristics to this PCEARP problem that outperform previous methods developed (Fig. 11.10).

It is important to refer that the household service is normally divided in the USA into "high density" areas, where thousands of addresses are georeferenced and in some cases are controlled by RFID instruments that indicate their filling percentages and other status. This helps districting, PCARP routing planning and following the

Fig. 11.10 Waste collection in bins (side-truck collection) – door-to-door bin system

execution process by IoT technology that supports such operation in a hole [31, 45]. The software used in this country, like as Opitrak™, Rapidis™, and FleetRoute™, [101, 102, 108], are applying the methodology of districting and routing for high density areas in general which are qualified to react with the changes in the field like new/removed houses or buildings/condominiums, exchanges in the network street constraints, interruptions, and others to facilitate drivers assistance and coverage also using web and mobile services.

Sniezek et al. [107] and later [106] considered a problem called a CARP with vehicle site dependencies (CARP-VSD) that appear in Philadelphia domiciliary waste collection. Considering a heterogeneous fleet of waste collection vehicles, a vehicle/site dependency on an arc exists if vehicles from one or more vehicle classes cannot service or traverse the arc while vehicles from other vehicle classes can service or traverse the arc. The vehicle/site dependency on an arc determines the vehicle classes that can:

1. Service the arc,
2. Travel along the arc but not service the arc, and
3. Neither travel along the arc nor service the arc.

Vehicle/site dependencies occur in residential solid waste collection due to such factors as some streets being too narrow for vehicles from certain vehicle classes to traverse or service and some bridges or overpasses not able to support the weight of the vehicles from some vehicle classes. The implementation of the solution brought to the city balanced and non-interlacing partition. The acceptability of the routes by the crew is a remarkable point addressed as the routes were more effective than existing ones (better balanced and with less intersections). The method is implemented in the RouteSmart™ software.

Canada

In Ontario, CA, cities are introducing split-body trucks that let the user load regular household garbage on one side and organic waste on the other. This improves the efficiency of collection, so instead of passing up and down the street multiple times, it can get everything all at once. Both have a distinctive "pendulum packer technology" that continuously packs the organics during the collection into a watertight body. The driver operates the pendulum from inside the cab, and it is constantly packing until switched off. The hopper is self-cleaning, and all the hydraulics are on the exterior of the truck for ease in maintenance. Options offered by the manufacturer include collecting with the side-lifting arm, which requires manual loading, or the automated lifting arm resembling a vertical conveyed belt. These provide collection functionality that range from manual to semiautomated to totally automated. These trucks are in use across Canada for residential organics collection, including in Quebec City (Quebec) and Barrie, Ontario.

Moazzeni et al. [83] considered the operation with electric vehicles to collect and transport solid waste, recyclables, and organics to recycling centers for transformation and disposal facilities. The authors developed a mathematical optimization model to coordinate decisions related to the location of waste collection centers and charging facilities with the optimal routing of vehicles and energy consumption objectives in pandemic outbreaks (like COVID). For this purpose, the mathematical optimization model considers minimizing the establishment cost of various facilities, the operational costs of waste collection activities subject to facility location, arc routing, and charge-calculation constraints. They also propose an efficient GA metaheuristic and a grey wolf optimizer (GWO) to obtain reliable solutions in real-world dimensions.

The software used in Canada, for waste collection route planning, among others, is Opitrak^{TM}. Opitrak is a Canadian company that delivers waste collection plans, following districting and routing services for high density areas where different types of bins are collected, as in US previously reported.

11.3.2 DWC in Central America

Mexico

Betanzo-Quezada et al. [12] investigated the efficiency of 71 domiciliary waste collection sectors/routes in the municipality of Santiago de Querétano, Mexico (626K inhabitants) using GPS and monitoring systems. By comparing the planned versus executed sectors, Querétano is one of the few municipalities in Mexico that RSU management is part of the municipality development plan, urban improvement, and major public health. It counts with a USW prevention and management specific law. The municipality cleaning was internationally recognizing.

The planned routes are designed according to the parameters related to the rate of total local population P and its population density d, (P/d), that must be close to $(a\ T\ r)/60$ where a is the proportion between the productive and unproductive distance of the route, the T time required to perform the route, and r is the vehicle velocity while in collection. No routing algorithms were used by the authors but only the parameters for the defined routes planned by the local manager by hand. No information was reveled about how the collection is done (bin or continuous door-to-door), but the study presented indicators of relative efficiency and debts between planned and real routes performed by the municipality.

Martinez et al. [81] proposed MILP models to improve planning of the design of territories for the cleaning, weeding, and collection of solid waste in municMartinez et al. ipal areas of San Nicolas de los Garza (476K inhabitants) public parks. Their study was evaluated in the municipality of the Monterrey Metropolitan Area, which considers factors like amount of waste and sectors visiting frequency. The authors define "macro-routing" as the allocation of collection vehicles to various areas of the city to perform the collection, and "micro-routing" as the specific route that must be met by daily collection vehicles in the areas of population where they have been assigned. The work developed two macro-routing MILP models, consisting of the optimal grouping of green spaces in the Monterrey Metropolitan Area, that create preferably balanced territories regarding the workload of each provider and the coverage of the collection service. In the case of space constraints, there is a basic set of conditions present in most applications that force the creation of contiguous, connected, and compact areas. As a result, all the parks of the municipality would be weeded once a month to generate a greater control in the allocation of parks to providers considering the payload.

Benitez-Bravo et al. [11] proposed a strategy for designing urban solid waste collection routes taking as baseline scenarios of collection routes in the Monterrey Metropolitan Area (San Pedro Garza García and Benito Juárez), with the least environmental and economic impacts through an optimization strategy that considers the road network's real-time data, added to an environmental analysis developed with a life cycle assessment approach.

The routes are performed door to door, but the authors used the best route obtained from TSP-based algorithms (with **R Studio**) with the center of the street segments to represent collection points. The results of the planned routes presented are quite notorious (20% reduction). The cost per route is estimated by the equation $TC_r = D_r\,(MC + CFC) + S_r,\ r \in R$, where TC_r is the total cost of a route r, D_r is the distance traveled by a route r in Km, MC is the maintenance cost per Km, and CFC is the cost of fuel consumption and S_r is the crew salary per route. The impact of the routes in the evaluated zones with different ratings in one-way and two-way streets results that as the number of two-way street increases in relation of one way, the major is the impact on savings in routing.

The study also investigated the impact of the traffic in the velocity of the collection, and it was clear the increase of 17.2% in waste collection velocity during the night in relation to the routes performed in the morning and in the afternoon. Considering environmental impact measured by the Carbon Credits, the authors reported that the simulations with the new routes revealed 6 and 9% in savings. By performing better routing schedules at night, the environmental impact is higher, because it affects the diesel consumption and efficiency.

Carmen-Nino et al. [27] considered aspects of a quantitative and qualitative research about waste management and routing waste collection in the presence of an unreliable waste management in the touristic city of Xaltianguis, Guerrero/MX (6.9K inhabitants). Xaltianguis is considered a locality with a low social backwardness index. The study used a GPS to identify 312 waste collection points in the city, and ArcGIS with its network analysis extension to analyze the operation of formal and informal routes. Formal routes perform a mixed collection of solid waste and combine the methods of fixed stops in secondary streets and sidewalks on main streets (door to door and bins).

The side-loading truck without compaction with a capacity of 6 tons is used on a single established route. The staff consists of a driver and one or two laborers who are tipped USD 0.25–0.50 by families, and they collect 2.8 ton/day at 60 points normally. An important point is that in high seasons, the collection is in double or triple per week. The municipal waste collection disposes the garbage in the closest landfill. Informal routes (4), on the other side, were done by locals independently with pickups-trucks that carry from (2.95 m^3) 591 up to (4.93 m^3) 988 kg/day in total of 5.1 ton/day. The collection frequency is daily (thee routes), but one of them is twice a week. The service fee per estate was different for each route, ranging from USD 0.50–1. The proposed optimized municipal routes by the study are a little longer than the executed routes, but they cover 32 new stops and performed them in 2 h less time (22% reduction). The new coverage reduced the daily visits to three times a week with a better coverage. The study used TSP-based algorithms from ArcGIS network analyst.

11.3.3 DWC in South America

Argentina

Early work on planning waste collection was proposed by Bonomo et al. [19] for the south area of Buenos Aires (2.89M inhabitants). The authors optimized the routes of waste collection vehicles servicing dumpster or skip-type containers (1 m^3), by reducing the waste collection just applying the classic traveling salesmen problem (TSP) asymmetric version, over the four collection sectors (subzones) evaluated. The sectors (subzones) were reviewed by the solution methodology using a naive subdivision but take into account the sequential positioning of the containers by their labels (consecutive labels means adjacent containers). The evaluation takes into consideration the impact of reducing the trucks' capacity. The application of the proposed method minimized the distance traveled by each collection vehicle in the areas studied, with reductions ranging from 10 to 40% of the existing routes.

Maheo et al. [78] applied a Benders decomposition method to deal with the domiciliary waste collection routing in bins, in Bahía Blanca, Buenos Aires (292K inhabitants) an important educational, industrial, and commercial center. They applied two of the tactical planning problems in this reverse logistic system with this approach: determining the location and/or capacity of garbage accumulation points and the design and schedule of collection routes for vehicles. They focus on a less traditional MSW design called garbage accumulation points (GAPs). A GAP is a predefined location in an urban area in which bins can be installed. Instead of providing a "door-to-door" pickup of garbage, constituents must drop their garbage at specific facility – the GAPs. These facilities can range from collective community bins to recycling centers. To support their methodology of using GAPs, they cite [21] and [52], where it is indicated that the use of GAPs reduces distance traveled while also leading to a smaller environmental impact through the reduction of greenhouse gas (GHG) emissions and other air pollution-related metrics.

The authors simulated instances using data from [28] and provide optimal allocation and routing for a variety of scenarios. In Bahía Blanca, the service is door to door, but in bins using a rear-loading truck (bins of 1.1, 2.4 or 3.2 m^3), the local authority wanted to move to the new GAP system by using a side-loading truck. The tests were performed to evaluate their Benders B&C approach comparing with MILP models both using or not valid inequalities (VIs) in solving instances. No comparison was made between actual system and new optimized system to sustain the arguments about door-to-door routing vs. GAP location and routing.

Greco et al. [56] developed for the city of Bahía Blanca the sequencing of micro-routes which is modeled as a vehicle routing problem with travel time limit and the vehicle's capacity. The authors proposed two MILP formulations and solve a set of instances of the city. They use this data to analyze the possibility of installing a new transfer station. At that time, the city waste collection was carried out with a door-to-door system in which the vehicle that collects the waste must visit all the dwellings of the city every day of the week except on Sunday. Vehicles carry the waste from the dwelling directly to the landfill, which is located outside the city, and which travel takes in average 45 min (34 km long).

This work addressed MSW collection from the perspective of considering fixed micro-routes. This is not a common issue in the related literature, but it has certain benefits such as not largely affecting the work routine of drivers and practitioners, which can be reluctant to drastic changes, and reduces the complexity of the problem allowing to address the routing plan of a whole city in a unique model, see also [13].

The authors define a *micro-route* as a relatively compact area in which the collection is performed by the same vehicle without intermediate stops. The city of Bahia Blanca has 32 micro-routes 17-day shift and 15 on night shift. The detailed internal path of the vehicle inside the micro-route area is already established, and the internal route of a micro-route is quite optimized in terms of distance and travel times given the experience of the personnel in charge. In this work, the micro-routes are considered as fixed and the problem addressed to improve collection logistics is their sequencing, that is, which micro-routes and in which order are visited by each collection vehicle. Then, the models consider visiting the landfill more than once, in order to unload and visit more micro-routes. To generate paths, the authors used OSRM (Open-Source Routing Machine – https://project-osrm org/). The micro-routes range from 5.4 to 11.2-ton, and the compactor supports 15.75-ton.

The study evaluated the inclusion of a new transfer station between two proposed places and identified a reduction of 5.25% in the traveling distance and three vehicles less. There are no other comments about costs of installing, operating, and maintenance of a new transfer station, although it is mentioned the important impacts of its installation as:

1. The collection vehicles spend less time traveling and more time in collecting waste.

Fig. 11.11 Door-to-door manual waste collection in Brazil

2. The personnel are more effectively employed, since more time is spent on collection.
3. The response rate to unexpected service emergencies from the collection vehicles is reduced, since vehicles are usually near to the area in which they operate.
4. Depending on the level of technification, transfer stations can apply some processing to the MSW such as waste sorting and classification which can reduce the material that is finally sent to the destination.
5. Roads to landfill or processing plants are less congested as one transfer vehicle may replace several collection vehicles (since increasing the size of the vehicle creates less congestion than increasing the number of vehicles).
6. Transfer stations can contribute to a reduction in pollution impact of the system and the quality of service provided to the citizens.

Brazil
As far as we know, the routing and scheduling studies in domiciliary waste collection in Brazil started at the universities headed by Federal of Rio de Janeiro at the COPPE-Production Department, with [79] and [80], where it was introduced the Eulerian solutions using a capacitated heuristic with the direct version of the Chinese Postman Problem (CPP) for an area of the Botafogo's borough. Later, [88], from COPPE-System Engineering Department proposed a node/arc routing heuristic solution method (Marcolix) for routing sectors (BAs) at worldwide famous Rio boroughs - Copacabana, Botafogo, and Urca performed by COMLURB, the municipality operator. The authors evaluated their routing plans in the field with savings from 8–13% in distance and more than 2 h per sector in reductions of the workload on lighter days of the alternate collection.

In Brazil, the majority of the cities employ continuous door-to-door collection, Fig. 11.11. The regulation of the system considers the income diversity of the population, and the use of bins is not common, but it is of frequent use for condominiums, great generators, and special collection points.

Lima et al. [75] investigated the application of the arc routing rural postman method implemented in Transcad™ [25] to evaluate routes performed by the concessionary that operates in the Itajubá/MG city (93K inhabitants). The authors considered the same sectors used and the internal divisions in circuits and compared the routes one by one. As result, the solution returned for the ten sectors evaluated with maximum of two trips per sector, a reduction of 18,49% in the total distance traveled and an increase of 3.87% in the total time spent by the fleet in the whole process is reported as final results.

Jacinto et al. [66] proposed a heuristic methodology indicated as a Multi-Capacitated Chinese Postman Problem (MCCPP) to solve DWC for 176 boroughs of the city of Cariacica/ES (383K inhabitants). The municipality desired to define a new partition for the city that increase the frequency of the coverage of the whole city but without changing the number of vehicles already in use. The authors developed a heuristic procedure to define such partitioning into routes up to 10-ton that maintains workloads up to 6 h and 15 min rest time per trip. The method takes the route and partitioning of sectors developed by Transcad^{TM}, not revealed by the manufacturer, CALIPER, and divide them in groups; the groups of sectors are distributed along three shifts daily along a week that maintain alternate (minimum of 48 h) the time between two visits to the same group. As result, the methodology returns 26 new travels weekly and 9.6% in improved capacity of the fleet.

In [121] and [122], Vecchi et al. proposed for the downtown city of Campo do Mourão/PR (99K inhabitants) a methodology to solve the DWC problem using a three-phase MILP modeling process. In the first step, with the oriented and weighted downtown graph, the garage as starting vertex and the ending vertex at the disposal site, the method solves the classical capacitated p-median problem (CpMP) where the capacity is related to the vehicles maximum load and the load is considered for each arc with demand (required). The medians and the related vertices define the coverage areas of each vehicle. In the second step, they solve an oriented open CARP model (OCARP) for each group generated, and finally the third step applies to the multigraphs generated by the Hierholzer's method to describe the Euler paths. As result, the authors applied the method for DWC and SWC, both evaluating the size of the fleet to best cover the area. Economies of less than 2% in DWC and 8% for SWC in travel distances were obtained in relation to the practiced by the concessionary with the solution.

Negreiros et al. [92] showed a DWC-DSS system, composed by sectoring and/or routing OR tools over a ESRI^{TM}-GIS MapObjects environment, as previously indicated. The DSS, called SisRot® LIX, is composed by tools for interactively graph-based modeling (the street network), including constraints as turn prohibitions, preference criteria for turn and/or service, and tools to assign people and apps to follow the execution of the planned routes while happening. In the DSS, a collection of heterogeneous capacitated clustering methods is composed to design the hierarchy of areas to be covered (zones, areas of the same frequency; sectors, areas where a vehicle works in a shift; and circuits of sectors, areas where vehicles perform a trip from garage to disposal site or form and to the disposal site).

The authors perform different types of sectoring and routing projects using the four mentioning basic strategies to solve the DWC problem by using clustering and routing alternatives with the Generalized Mixed Rural Postman (GMRPP) approach to design circuits' routes, Fig. 11.3. Capacitated clustering methods used were from [92], and the GMRPP methods were from [93]. As result, the authors reported different uses of the methodology in DWC for five downtown sectors of the city Campo Grande/MS (898K inhabitants), the same set for the human street sweeping for sectors of the sub-municipally of Butantã in São Paulo city (11.4M inhabitants), and for the mechanical street sweeping with sweep tankers for sectors of Campo Limpo sub-municipality. As results, the methodology returns savings up to 28% in distance and reduced a vehicle and a crew (driver and three laborers) against the fifth units used, for Campo Grande/MS and reduced mostly in 35% the distances of the routes performed by the human sweepers for Butantã sectors evaluated and reduced a sector, distance and time to perform the coverage of Campo Limpo sub-municipality, up to 33% in savings.

Negreiros et al. [93] reported a new set of successful implementations for DWC and SWC, for the downtown city of Petrópolis/RJ (279K inhabitants) and Bom Jesus dos Perdões/SP (21.9K inhabitants). The authors present new tools for the DSS to follow the projected routes in the new designed sectors and revealed in a full the experience they had in design sector-routing projects for waste collection in Brazilian cities, as the tests of execution of their methodology in the city of Petrópolis/RJ. As result in the DWC, the five sectors were reduced to four, and the distance traveled was reduced in 12%, meaning a total reduction of 17% with fleet, variable and crew's costs. The sectors and routes were tested and implemented by the concessionary.

Negreiros et al. [95] proposed a new arc routing problem called he Twin-Rural Postman Problem (Twin-RPP) to deal with the domiciliary waste collection in the cities of Andradina/SP (57.2K inhabitants), Jales/SP (49.2K inhabitants), and Santa Cruz do Rio Pardo/SP (47.9K inhabitants). The Twin-RPP can be considered a pure arc routing based problem, with similarities to the Close Enough Arc Routing Problem (CEARP), [39]. These are cities in the country (rural area) of the state of São Paulo/Brazil. Due to the low budget to the domiciliary waste collection, these cities agree the manual collection by laborers without the presence of the vehicle door to door and the accumulation of garbage in corners for 20–60 min, Fig. 11.12. This means that the domiciliary service is composed in two parts: by the vehicle with laborers in movement and only by laborers in some streets. The synchronization of the services and the high effort of the laborers while walking and doing manual collection are the major challenges of this problem.

The Twin-RPP can be considered a pure arc routing based problem, with similarities to the Close Enough Arc Routing Problem (CEARP), [39]. This problem was considered as a multi-objective problem and a column generation procedure was designed to produce the schedules. As results, the authors executed their methodology in Andradina where the solution provided approximately 35% in less distance performed by the manual laborers, and 23.07% reduction in the total travel distances between sectors weekly performed by the fleet.

In Fig. 11.12, it can be the process mentioned above, of garbage accumulation in corners and below the walking movement of two laborers, while the vehicle traverses the other streets of the sector, collecting in low march (<5.2 km/h) to have time to perform the waste collection without losing previously accumulated garbage in corners. Images of the routes obtained with CherryTrack® LIX.

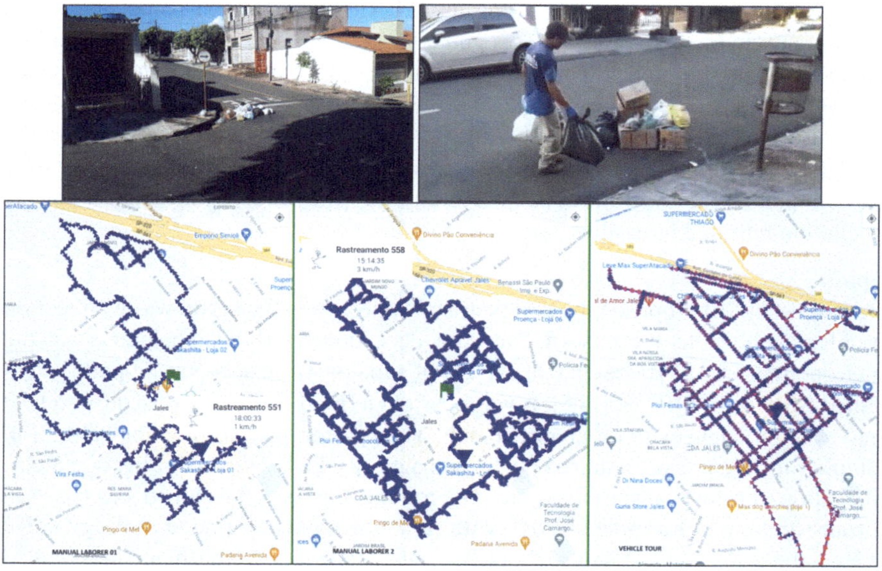

Fig. 11.12 Door-to-door waste collection with temporary accumulation in comers

11.3.4 DWC in Europe

Italy

Ghiani et al. [48] reported a cluster-first-route-second approach applied to a small town of Castrovillari/Italy (25K inhabitants) for the Health Department. The research focus was on residential sanitation routes that were formed manually by the local supervisor. Main problem related was building the transit network, because it was founded in Middle Ages and most streets are narrow and can only be traversed by small vehicles. To avoid traffic congestion, some places need time windows specified by local authorities, and some vehicles are not able to unload some containers. The collection was done by three specially equipped trucks of different capacities and features. The vehicles depart from the garage (at downtown) and dump in a sanitary landfill located in the outskirts of the town. The collection is made in containers of two types (1.1–285 units and 1.7 m^3–100 units). Both types of containers can be unloaded by two of the vehicles but not by the smaller. The waste collection had to be performed from Monday to Saturday from 5:00 to 11:00 shifts, with a crew of two workers and one driver. To avoid congestion, the main streets were divided into three classes, where each class has to be finalized in a specific deadline.

The health department wanted to (a) minimize route cost closest to optimal; (b) overtime could be eliminated, and (c) reduce de number of vehicles. To deal with the problem, the authors developed a cluster first route second approach as follows: They choose as clustering a p-median problem (adjusting assignment costs and types of services) and as routing an MCARP-IF approach respecting links' classes and divided the required arcs and edges (LR) as the ones with at least one container on its side and classified them into three (LT). Travel (30 km/h) and service unloading times (80s) were associated with each link (L) and required link (LR), respectively. The proposed two-phase heuristic returned a savings of 7.67% in the total cost, corresponding in 10.6% in distance and 13.5% reduction on working time and eliminated the overtime. They tried to reduce the number of vehicles, but the resolution returned infeasible.

Spain
Bautista et al. [6] proposed the first heuristic approach to the urban waste collection problem applied to the city of Sant Boi de Llobregat (80K inhabitants) in the metropolitan area of Barcelona that is part of a major waste decision support system called SIRUS. In SIRUS, [115], it was possible to modify routes constantly to take into account changes in urban circulation. The system provides a graphical tool to simplify different tasks required by the managers as the location of collection areas, allocation of refuse bins, routing, etc. The routing approach developed was based in a transformation of a mixed CARP with turn constraints to a node routing GVRP (Generalized Vehicle Routing Problem), describing a new problem they called Mixed Capacitated Arc Routing Problem with Turn Constraints (MCARPTC).

The application in the city considers the use of one or more refuse bins (1 to 3.2 m^3) at each collection points where the municipality living close dispose their garbage. The bins are located on the sidewalks next to the roads in designated areas surrounded by a small border. The city contains 1316 forbidden turns. Most forbidden turns (1008) are U-turns, whereas 308 forbidden turns correspond to turns that "connect" two different arcs/edges. Metrics for waste production per link was used by the population density. The capacity of the five vehicles used was the same (25 m^3) with a compact factor of 1/6. The authors developed two Ant Colony-based metaheuristic adjusted to the problem and tested their results in comparison to the practiced five routes developed. The resulting routes that they obtain suppose a reduction on the total length of at most 35–37%, meaning a reduction of 22,000 km per year to collect waste, impacting in the reduction of fuel consumption, environmental benefit due to gases emissions reduction, and hours of labor that could be transferred to maintenance and sweeping of streets.

Portugal
Ghiani et al. [50] proposed a metaheuristic to optimize routes of the refuse collection vehicles in Seixal (150K inhabitants), one of the 18 municipalities of the metropolitan area of Lisbon in Portugal. Two models were used and solved by CPLEX, the first obtain a valid lower bound and the second a valid solution to MCARP. They are connected by a heuristic method to fix values of some variables (i.e., services that are assigned to specific trips). As a subproblem is solved, where the number of variables is decreased dramatically, the exact model produces feasible solutions.

The Seixal Waste Division established some zones to evaluate the proposed methods. The parameters of the required network were collected from the historical data from past years (amount of waste per street, the service, and deadheading times). The constraints of maximum load and workload were imposed. The method was applied to Seixal networks of small size (up to 361 vertices and 713 links) taking long running times and far gaps from LB/UB, but the returning solutions were considered "as good" by end users.

Bianchi-Aguiar et al. [14] proposed the use of a Periodic Vehicle Routing Problem (PVRP) approach to solve the waste collection problem integrated with the scheduling of visiting residential containers of the city Ponte de Lima (43.5K inhabitants), Portugal. In Ponte de Lima, collection plans were still hand-made. The municipality manages the collection of the waste generated in the region where they own a set of waste collection vehicles located in a single garage (depot). Containers are visited by the collection vehicles a specific number of times in a week, negotiated with the population. The collection frequency differs among containers. In any case, the population complained about the excessive quantity of waste accumulated and the regularity of the service. The objective of the study was the reduction of the collection costs and secondly to reduce the dissatisfaction of the inhabitants. As output, they proposed a visiting calendar for each container within the weekly planning horizon, and a 1-week plan of daily routes for each collection vehicle.

The authors carefully evaluated container production and the routes executed, to design their maps, and developed a hierarchical approach of methods to solve the problem proposed. As results, the methodology obtained a reduction of 26% in the distance performed by the vehicles during collection, although the dissatisfaction of the inhabitants was also tackled in the delivered plans. But the best interests of the employees are achieved, because it is guaranteed that the work shift duration was not exceeded.

Finland

Nuortio et al. [97] described the optimization of vehicle routes and schedules for collecting municipal solid waste in Eastern Finland. The collection process is made in locations of bins that can be aggregated in macro-bins to facilitate waste removal, the departure and destination point are the same, and the fleet is heterogeneous, and regulations on crew worktime and service locations were defined for the period length set to 4 weeks, and most of the containers have 7 or 14 days as their schedule interval, i.e., the containers must be collected once a week or every 2 weeks for scheduling. The solutions are generated by guided variable neighborhood thresholding metaheuristic that was adapted to solve the proposed real-life waste collection problem.

A case study on the waste collection in two regions of Eastern Finland demonstrates significant route improvements are quickly achieved also because of combined routing and scheduling optimization methodology employed. Average saving in the city of Kuopio with the routes indicated an improvement by nearly 46%, for the area of the city of Pieksämäki, although the data on the visits to disposal site (transfer station) was missing than these problems hamper the comparison of the executed and the optimized schedules.

Greece

Chalkias and Lisaridi [29, 30] reported an optimization methodology for waste collection based on GIS real data of the Municipality of Nikea, Athens, Greece. The strategy consisted of replacing and reallocating the waste collection bins as well as rescheduling waste collection via GIS routing optimization. The benefits of the proposed strategy were assessed in terms of minimizing collection time, distance traveled and man effort, and, consequently, financial and environmental costs of the collection system. Nikea (MoN) is a typical Greek urban municipality (96K inhabitants, producing 42-ton/year and territorial distributed in 6.65 km^2), characterized by multistory apartment buildings, combined by lower multiple dwellings (two to four apartments) and mixed residential and commercial land uses in many neighborhoods. The collection system was made by 17 rear-end compaction trucks with an average of 9-ton (in a heterogeneous fleet) that cover approximately 12,000 wheelie bins. The crew size on the vehicle is three persons, a driver who never leaves the truck (as required by safety regulations) and two workers who move and align the bins with the hydraulic lifting mechanism of the truck.

The municipality is empirically divided into 15 sectors (collection zones), each of which is further divided into 2 subsectors. Waste is collected in each subsector every other day. Most of the bins are small, of 120 and 240 L capacity, but a few larger ones exist in some central points. In total, there are currently 12,107 bins in MoN, of which 875 of 120 L, 9526 of 240 L, 504 of 330 L, 1110 of 660 L, and 92 of 1200 L. Due to the large number of small bins, it is not possible to make a full mechanical collection. Additionally to every sector, a new worker is sent starting its shift one hour earlier than the vehicle. This worker either moves small bins from their location to central points and crossroads, from where they will be mechanically collected, or manually empties the content of bins with low load to other more central ones, which in turn are mechanically collected. In total, it is estimated that only 70% of the bins are mechanically collected, with the content of the rest being manually transferred in other bins. The worker going ahead on foot and the two workers on the vehicle perform their duties in rotation.

The municipality wanted to reduce the number of small bins to larger (1.1 m^3), and the problem turns to a location-routing problem. The authors performed an spatial analysis of the small bins and their production, and they select locations as follows:

a. Allocate bins on the road network (intersections are preferable);
b. Install new bins near existing bin locations (in a buffer zone 60 m wide); and (c) allow the placement of more than one bin at the same intersection. The number of the bins sharing the same intersection point is related to the surrounding land use and the population of the covered area.

Using ArcGIS Network Analyst, they performed the location of the new bins and the routing steps with an Asymmetrical TSP method. Thus, a maximum travel distance of 60 m from each resident to the proposed new site of the bin was allowed. They verified their method by comparing the actual system with the new reallocation of bins, and they obtained 12% in savings on distance traveled and time savings of traveling and emptying of 17.5% inside the sector.

11.3.5 DWC in Africa

Nigeria

Ogwueleka [99] showed that in Nigeria, solid waste collection vehicles are assigned to zones without an appropriate focus on waste generation analysis, and routes are left to the drivers. Every time the vehicle is filled up, it heads to the disposal site to unload and then returns to the zones. As it is done without an appropriate methodology, this contributes to high solid waste collection cost. The study considered a symmetric CARP approach that was applied to the city of Onitsha (509.5K inhabitants). Municipality is subdivided into six zones: Okpoko, Fegge, Housing Estate, Upper Iweka, Inland, and GRA.

The study concentrated in the district of GRA, a major residential area, whose urbanistic structure was similar to the whole city. The local waste collection administration adopted hauled container system (HSC) and stationary container system (SCS). It was common to the vehicles operated above full capacity (5–10%); there was a disposal site, and they performed two trips for each sector, a daily journey of 8 h with a driver and four laborers per sector. To apply the methodology, the author conducted an accurate measure of the waste collection production based on the vehicles' trips, network analysis, and positioning in the network the distribution of the containers along the area as it was properly identified.

As result of applying the CARP heuristic method, the executed routes with a 5 m^3 truck were reduced from 5 to 4 vehicles, meaning savings of 16,3% in distance, 25.24% in collection cost and 23.51% in collection time, if the appropriate truck was applied. After evaluating three types of them (5, 7.5, and 15 m^3), the results went favorably to the trucks of 15 m^3, although the collection cost is higher (27,85% more, from the 5 m^3) the incomparable cost of hauling (60% less, from the 5 m^3) make a final reduction of 27.82% in the final total cost.

Algeria

Abdelli et al. [1] proposed an optimization methodology for minimizing both logistics and environmental impact by gases emissions. The classic vehicle routing problem (VRP) was used (distance and time traveled), and the routing system for household wastes collection based on the existing network of containers, the capacity of vehicles, and the quantities generated in every collecting point.

The work was done in the city of Mostaganem one of the largest cities of Western Algeria (180K inhabitants) that produced about 130 tons of household solid wastes a day. The city was divided in 16 sectors, with an average length of the collection route of 155.5 km. Trucks from 8 to 12 m^3 of waste were used by the municipality, and the cost of collection was about 18.7 Euros/ton in 2014. They used ArcGis 10.2 software and the Analyst Network extension (ESRI) series to optimize the household waste collection.

The authors evaluated different scenarios:

S1: Route optimization;
S2: Time optimization;
S3: Route optimization with a proposal to use a 12 m^3 truck capacity;
S4: Route optimization with a proposal to construct a waste transfer station.

The study was performed over total costs of collection (including workforce, equipment, etc.) and gases emissions with relative equations to travel distances and time.

As result, S1 returned a reduction of 42.8% from practiced total cost and pollutant emissions reduced 6.9%; S2 returned the total cost of the collection has been reduced by 42.2% and pollutant emissions reduced 1.8%; S3, returned the total cost of the collection has been reduced by 44.9% and pollutant emissions reduced 34.8%; and S4 returned the total cost of the collection has been reduced by 46.5% and pollutant emissions reduced 67.9%.

Tunisia

Kallel et al. [67] studied optimized scenarios with ArcGIS Network Analyst in order to improve the efficiency of waste collection and transportation in the district Cité El Habib of Sfax city, Tunisia (330K inhabitants). The scenarios were:

S1: Route optimized with the same working resources (change of stops sequencing only);
S2: Route optimized with change of vehicles; and
S3: Route optimized with change of collection method (vehicles and reallocation of bins).

A result showed the three scenarios guarantee savings compared to the practiced in terms of collection time (14%, 57%, and 57% for S1, S2, and S3, respectively) and distance (13.5%, 13.5%, and 40.5%, respectively).

The work presents the distribution of the collection costs of the districts between staff (62%), equipment (10%), subcontracting (9%), waste transfer (5%), depreciation (12%), other costs (2%), and total costs TND/ton and TND/cap (TND-Tunisian Dinar). The study collected information about routes' location of the starting point, starting time, number of workers, itinerary and coordinates of collection points, condition of waste on site, condition of container and bins, odometer reading before departure and after collection, time of arrival at the transfer station, amount of waste collected, and quantity of energy consumed.

With ArcGIS Network Analyst, they observed their results over the traffic network built for the city, by solving ATSP between bins of the selected areas, as other previous works. The last scenario solution, which basically replace anarchic points to containers placed at most 125 m to the waste producers, indicates that its application would bring 14,700 US of annual savings to the city.

Table 11.2 Impact evaluation of choosing different type of vehicles to the waste collection system

	Truck type								
	6t			10t			18t		
Scenario	Trips	Km	Hours	Trips	Km	Hours	Trips	Km	Hours
S1 (Kiteesi)	82	1407	252	48	853	202	27	563	184
S2 (Kiteesi)	218	3687	436	125	2182	317	65	1219	240
S3 (Bukusuma)	216	6007	309	125	3547	247	65	1916	204
S3 (Wakiso)	211	5924	293	126	3493	239	66	1899	201

Uganda

Kinobe et al. [71] proposed a study on waste collection considering the temporary collection points that vary depending on waste deposited. If there is little waste at the temporary collection points, the trucks are directed to other pressing sites that have huge quantities of illegally dumped waste.

The study was carried out in all the five divisions of Kampala City district. The city has 99 parishes with a population of about 1.72 million people during nighttime and estimated to be generating 1200–1500 tons of waste per day. The waste generated took as reference the waste disposed in Kiteezi landfill that is about 40% of the total waste produced by the city. The ArcGis map software was used to define the georeferenced database, location of dumpsites, truck capacities, the road network, as well as the waste generation volumes of the area. The software calculated the distance and travel time of the truck per trip, with a constant speed parameter. The Network Analyst tool was used with the VRP algorithms to determine the shortest route from the collection points to the landfill in the same manner as it has been used in urban network analysis. The model used was the VRP with Time Windows (VRP-TW), where the fleet was considered homogeneous, the service time was the same for all parish points (20 min) as its load, and the travel speed was constant (30 km/h), although traffic conditions were considered according to the time of the day by the software. The routes' start-up point was the central division headquarters, and a constant collection is done until the truck is filled and departed to the landfill.

The authors evaluated three scenarios:

S1: Routes with 6-, 10-, and 18-ton trucks of 40% waste collection;
S2: Routes with 6-, 10-, and 18-ton trucks of 100% waste collection;
S3: Comparison of optimized routes using KCCA trucks and outsourced private operator using 20-ton truck at 100% waste collection to Kiteezi and proposed new landfill sites (Bukusuma, Kalagi, Semutu, Wakiso, Zirobwe).

Table 11.2 presents the results obtained with the scenario evaluation using ArcGis Network Analyst, using Keteesi landfill in scenarios 1 and 2 and Bakusuma e Wakiso in scenario 3.

The authors emphasize a great environmental benefit in using heavy trucks (18t) which is reducing greenhous gas (GHG) emissions and the reduction in costs of manpower, time, and operational costs during waste collection and transportation.

Ghana

Sulemana et al. [118] investigated the application of optimized routes for solid waste collection on Kwame Nkrumah University of Science and Technology (KNUST) campus in Kumasi, Ghana. The study modeled the existing routes for a complete collection cycle using travel time criteria and generated optimized routes also using ArcGIS Network Analyst with VRP. The validation of the optimized outcome (travel distance and travel time) was done by subjecting the solid waste collection trucks to the optimized routes over real traffic conditions and collecting bins of different natures (pegged, small, and communal). The results confirmed the advantage of using VRP to this approach with 14.5% of savings in total distance and 9% in total time spent in labor activities.

Odum et al. [98] studied ant-colony systems adjusted to solve capacitated clustering vehicle routing problem (CCVRP) to minimize costs of waste collection of Pankwase area of the city Sanyani/Ghana. The solid waste collection in the municipality is done by a concessionary limited at centralized locations and then transported finally to the landfill site using the following vehicles: skip, compactor, and roll-on-roll-off. However, owing to financial constraints, increase in fuel cost, high cost of labor and vehicle maintenance, and poor road network, some communities (as Penkwase) are usually not covered during the solid waste collection on daily basis.

The results shown are based on the capacitated clustering solution and the tour of the best ants – as in sector (node) routing. The solid waste collection group has two (2) trucks that are in good condition to operate on different routes the total cost of transporting the waste was reduced by approximately 50%.

11.3.6 DWC in Asia

Palestine

Assaf and Saleh [5] studied a point-to-point waste collection for the city of Nablus-Palestine (351.4K inhabitants) including the old city, villages, and three refugee camps) southern region, via a genetic algorithm applied to the VRP, to minimize total distance traveled and corresponding total incurring costs. With difficulty and creativity, the authors collected the data from the remaining dumpsters and introduced them in the ArcGis software. The waste data were collected and notice its great variability, and the problems caused by them (e.g., unfixed truck routing, some unvisited dumpsters are left to the next day, etc.).

Comparing to the executed service, the solver returned total traveled distance was reduced by 66%, whereas the collection time was reduced from 7 hours per truck-trip to 2.3 hours. The researchers presented the plans to the head of the public sectors, and the optimal routes were validated to be considered a realistic solution.

Iran

Madahvi et al. [77] studied a sustainable multi-period and multi-trip vehicle routing problem integrated with relocation models that were developed to redesign the intermediate transfer stations and find optimal vehicle routes and the best collection frequency for each municipal solid waste generation point. The waste collection system was organized in three levels: waste generation regions, intermediate transfer stations, and final destination (recycling centers, treatment facilities, disposal centers, landfill sites). In the waste generation centers, the collection is made in containers (59kg max capacity) that are located along the regions.

The model integrated the three levels of the logistic system once they operate in conjunction, but it can only solve small networks (tested up to 4100 inhabitants). The model was applied to the city of Tehran (8.6M inhabitants) as case study, where in their simulations, the results indicated that system social score was affected by collection frequency decisions, redesign policy, and the number of demand nodes. Other results came as total cost reduction of 66% that occurred by a 86% reduction in weekly traveled distance and a 12% decrease in routing social score.

Turkey

Düzgün et al. [42] proposed a CF-RS solution using a parallelized hybrid GA multi-objective (PHGA) method for waste collection on bins. In this case, the routes may not perform U-Turns, and the steep slope of the relief is considered for every movement of the trucks. The fuel consumption is considered using equations proposed by [119] that is a function that considers the accumulated distance, weight, and slope. Their results indicate that although one model produces longer paths compared with the other proposed, it prevents waste collection uphill when the collection truck is loaded.

This study introduces the software **RouteSW** for the optimization of MSW collection routes by considering path constraints. The point-to-point collection routes were evaluated in the Bahcelievler and Emek districts of Ankara/Turkey.

Vietnam

Son and Louati [116] proposed a generalized vehicle routing model including multiple transfer stations, gather sites, and heterogeneous fleet of vehicles with time windows for MSW collection. It takes notice of traveling in one-way routes, the number of vehicles per m^2, and waiting time at traffic stops for reduction of operational time. The objectives are to maximize the collected waste quantities and to minimize the environmental emissions (the impact of climate change).

Their case study was at Danang city (1.34M inhabitants), which is one of largest industrial centers of Vietnam. The problem was proposed in two parts:

- Part 1: Household (street) solid wastes are collected by handcarts and assembled at gather sites close. Each site has its own time window that varies waste quantities as they are different.

- Part 2: In each shift, a vehicle starts from the depot and moves through gather sites following by a scheduled route and finishes at the landfill.

This logistic system can be followed in Fig. 11.13, where the phases are identified, and also the preparation of the city to include the handcarts with new electric garage truck units.

When moving to a site, the vehicle loads garbage with a waiting time for this process. Along the route, once the vehicle is full of garbage, it moves to the landfill for unloading and complete a route. After reaching programmed routes in a tour, the vehicle comes back to the depot. The objective function aims to maximize the collected waste quantities and minimize the environmental emission with respect to vehicles. Their model was implemented with Phyton scripts for processing tasks, and they integrated it into ArcGIS Network Analyst. The authors evaluated different criteria (economic, environmental, technological, sociopolitical) and related sub-criteria (operation and maintenance costs and fuel cost; quantity of waste, fuel emission and noise; reliability and efficiency; labor acceptance and compatibility with the national energy policy objectives).

The experimental results showed that the total traveling distances and operational hours of vehicles of the new model are less than those of the practical routes. Environmental analysis using the AHP method was done for better investigation of the results which emphasized the applicability of the proposed measures to waste management.

Taking into account the previous work, Vu et al. [124] focused on the interrelationships between the parameters in a dual-phase collection system by explicitly considering scenarios with different numbers of TCPs (temporary collection points) and source-to-TCP distances. Their objective was to develop a simple GIS-based approach to minimize system costs considering both pre-collection and collection stages. The study was done in Hai Phong City (1.8M inhabitants) city center including five wards with a population of 28,450, and covering an area of 1.41 km^2. It also examines how the interrelationships between the number of TCPs and the maximum source-to-TCP distance affect the waste system's cost.

The study area included both pre-collection and collection stages, in daily collection routes to households, small shops, and institutions by handcrafts with capacity of 0.5 m^3, limited time to 8 h/day. The truck capacity used in collection phase was a 10 m^3 unit with daily travel of 120 km during 8 h/day. For the handcrafts, the study used the VRP module of ArcGIS Network Analyst to evaluate their routes with no turn, narrow alleys, and one-way prohibitions, but a walking speed of 3 km/h. After applying their model, and evaluated the different scenarios, the authors reported the recommended scenario helped to reduce truck travel distances by 13.76% compared to executed, and handcraft limited to a maximum distance of 500 m.

Fig. 11.13 Logistic waste collection domiciliary and selective system in Vietnam

Fig. 11.14 Example of a mobile depot in selective waste collection

11.4 Domiciliary Selective Waste Collection

The selective waste collection is performed in many different ways around the world. We selected some examples from Italy, the Czech Republic, Brazil, and Argentina to better illustrate these alternatives to process the service.

Italy

Del Pia and Filipe [41] presented a divergent strategy of selective waste collection (Paper – Plastic – Metal) proposed to the small city of Due Carrare (8.5K inhabitants). Due Carrare is a small town located in Province of Padova, Northern Italy. The problem consists in the need for arranging appointments between vehicles along their routes, so that small vehicles (called satellites) can dump their contents in the large ones (compactors) and continue their work.

They proposed a generalized CARP (CARP-Mobile Depots)-based algorithm, by using VND metaheuristic, that as a result decreases the total route duration of about 30% with respect to the routes so far adopted by the municipality (Fig. 11.14).

Caramia et al. [26] investigated a WMS to attend the municipality of Lazio, Italy. The authority wants to foresee the behavior of the concessionaries to avoid shortsighted decisions. In the first stage, the regional authority divides the clients into clusters, further assigning located sorting facilities to these clusters. In the second stage, an effective route was defined to serve clients. The main idea behind the model was that the authority aims to find the best location–allocation solution by clustering clients and assigning facilities to these clusters without generating overlaps. In doing so, the authority tried to (i) assign the demand of clients to the facilities by considering a safety stock within their capacities to avoid shortages during the operational phase, (ii) minimize greenhouse gas emissions, and (iii) be as compliant as possible with the solution found by the second-stage problem, the latter aiming at optimizing vehicle tour lengths.

The two-phase metaheuristic proposed (solving constrained facility location problem and a Tabu Search – TS, to correct assignments to customers and routes) explores the three aspects of decision planning: strategic (locating facilities), tactical by clustering customers accordingly to their behavior, and operational by performing routes between clusters and their assigned facility. As result, the authors applied their methodology with success for the municipally of Lazio, saving sorting stock levels and environmental impacts with most appropriate facilities selected.

The Czech Republic

Viktorin et al. [125] proposed a MILP model to locate waste bins and containers when optimizing the number of them and the walking distance from the waste producers to the assigned waste unit. Although their proposed MILP model does not apply to medium-large-sized cities, they propose a hierarchical clustering-based heuristic to simplify the model and deal with large-scale instances. Problems combining the decisions on location (where to locate) and allocation (size and number of bins to locate) are often called location allocation problems.

Reducing the number of bins and minimizing the distance between waste producer and its bin were considered conflicting objectives. Hierarchical clustering method (Ward's) was used to define the natural classes of waste producers due to its characteristics (residential area, industrial, old residential, downtown, etc.). The authors applied their metaheuristic for a set of artificial instances of a moderate sized (6212 address points after filtering) and after to two case studies: Zlín and Prague, to selective waste collection dealing with the allocation of cooking oils and fat waste containers. In the city of Zlín (51.7K inhabitants, Czech Republic), this city was selected because it featured a densely populated city center, large housing estates typical for Czech cities, industrial zone and old household streets, and remote populated satellite towns. Between the representative selection and the subproblem definition heuristic strategies, the representative selection approach provides relatively evenly distributed representatives after the clustering step, and thus it was able to provide a more evenly distributed collection infrastructure. This leads to easily implementable solutions for real-world scenarios. For the second case study in Prague (1324M inhabitants, Czech Republic), it involves more than 102,000 address points, and various areas characteristics. The hierarchical method returns 14 areas that were allocated 1328 containers to a maximum walking distance of 500 m.

Brazil

For the tactical planning, Simonetto and Borenstein [114] proposed a DSS for selective waste collection through generating alternatives to the decision processes concerning the allocation of separate collection vehicles as well as the determination of their routes and the determination of the daily amount of solid waste to be sent to each sorting unit in order to avoid waste of labor force and to reduce the amount of waste sent to the landfills. Later, Li et al. [73, 74] proposed a set of models and a heuristic method for a variation of the Single Depot Vehicle Scheduling Problem (SDVSP) for reducing the number of trucks that covers static trips already defined by managers of selective waste collection in the city of Porto Alegre/Brazil. In comparison with manual scheduling, the proposed method achieved 25.61% estimated savings of the operational costs, also improving social impacts in noisy and waste management by reducing daily trip assignments to recycling facilities.

Gil and Avila [51] analyzed the economic and environmental efficiency and impacts on urban traffic models for collection and transportation of recyclable materials door to door through a comparative study of three models running in Brazil, namely, truck compactor, box truck and motor vehicle combustion, unmanned, associated with a point of support. The work confirmed the thesis that proper planning for resource optimization, efficiency, and quality in service delivery is a key activity to the implementation and maintenance of selective collection programs, in a way to keep compliance with the National Policy on Solid Waste.

The authors evaluated their study in three different municipalities of Rio de Janeiro/RJ (COMLURB) with load packers, João Molevade/MG (Cooperativa Popular Amigos do Meio Ambiente) with tank vehicles and in the city of Jacobina/BA (Cooperativa de Catadores de Materiais Recicláveis Recicla Jacobina) with a fuel-based automatic vehicle.

Negreiros et al. [93] reported a project design to SWC, for the municipally of Bom Jesus dos Perdões/SP. The sector-routing methodology is explored, and mainly the GMRPP was in evidence, because it is shown the presence of vertices with demands – condominiums (required vertices), to treat difficulties related to the waste collection sectoring-routing problem in the presence of local garbage appeared in vacation/holidays season. They showed the detailed project defined to design complex routes for domiciliary selective waste collection.

Argentina

Braier et al. [22] proposed to optimize the routes of a recyclable waste collection system service at Morón, Argentibe. Morón (320K inhabitants) is a large municipality that was divided into five administrative districts (Castelar, Morón City, Haedo, Palomar, and Villa Sarmiento). The city is also subdivided into sectors that are covered in a specific day of the week (called green day) by a recycling truck that must traverse its sector in one trip.

The project objective was to maximize the level of service to not skip any street segment. The optimized open RPP routes imply a lower fuel consumption, reduced emissions, reduced collection times, and less traffic jams (since the collection trucks have a quite slow average traverse speed). Restrictions on performing services are the Argentinean Road traffic regulations prohibiting left turns at signal-controlled intersections as well as U-turns.

The authors consider the problem as a particular case of the generalized rural postman problem (gRPP), where given a directed graph $G = (V, A)$, a set of required arcs A_R, and a set of groups of nodes/arcs, the problem consists in finding a minimum-cost cycle traversing all the required arcs and at least one element from each group. The gRPP model formulated was applied to the five sectors individually and was implemented by the municipality. The final solution performed significantly better than the manually defined routes used previously. The main advance was that the final solution covered every street within the municipal limits. Before implementation, up to 60 street blocks for 1 sector and 16% of street blocks for another sector were skipped in certain sectors. This improvement was achieved while in some cases reducing total vehicle distance compared with the old manual routes. In few cases, distance did increase, but only slightly. After several months of testing the routes defined by the MILP methodology, the municipality began implementation in mid-2014. The success of the project has been such that the authors were invited by Argentina's Ministry of the Interior and Transport to replicate the experience in other municipalities around the country.

11.5 Challenges in Industrial, Commercial, and Health Waste Collection

The operational problem on waste collection for industrial, commercial, and health waste collection diverges as in the case of residential waste collection.

Industrial waste collections are generally made in big containers ($3, \ldots, 10\,\mathrm{m}^3$) needing services with roll-on-roll-off trucks.

In commercial waste collection the service is made in bins and containers ($0.2, \ldots, 2\,\mathrm{m}^3$). A diverse number of strategies can be found, as simply capacitated and time constrained open node routing (basic VRP), VRP with time windows, periodic VRP with time windows, and node sector routing with time windows.

Finally, in health waste collection, the service is done in appropriate bins ($0.1, \ldots, 0.6\,\mathrm{m}^3$) and the problem considers the relevant aspects related to the hazardous materials collected. In this case, it is important to take care of the customer waste production and periodicity, as the aspects of routing in three phases (garage–customers–incinerator, incinerator–customer–incinerator, and finally incinerator–garage). Below we followed some real applications obtained by this study.

Hauge et al. [62] presented a practical roll-on/roll-off routing (ROROR) problem arising in the collection of industrial waste. It deals with transportation of bulky waste containers. Skip containers, which are used for the waste collection, need to be distributed between, and collected from, a set of customers. Full containers must be driven to dump sites, while empty containers must be returned to the depot to await further assignments. Unlike the traditional ROROR problem, where vehicles may transport one skip container at a time regardless of whether it is full or not, the authors considered cases in which a vehicle can transport up to eight containers, at most two of which can be full.

The authors investigated a set partitioning model using a hybrid column generation approach (with TS as a fast column generator) and a pure Tabu Search metaheuristic to obtain their solution to the problem. The hybrid column generation obtained the best results.

Sahoo et al. [104] developed a node routing system called **WasteRoute** to Waste Management, Inc. Waste Management is a company that operates with commercial waste collection. According to the authors, WasteRoute is a comprehensive route-management system that considered WM's specific routing concerns (route compactness and workload balancing) and provided broad benefit. WM deployed WasteRoute across the USA beginning in the early 2003. By the end of 2003, WM had 984 fewer routes, saving 18 million USD. Kim et al. [70] presented more carefully the methodology employed by WasteRoute that uses a capacitated clustering-based waste collection VRPTW with multiple depots and multiple intermediate facilities. The methodology is a cluster first and route second with time windows that uses for capacitated clustering a modified capacitated k-means and Solomon's insertion method for time windows.

Negreiros et al. [93] applied a periodic capacitated clustering problem to design clusters related to different vehicle types for the health waste collection in Fortaleza (2.3M inhabitants), Brazil. The health units were carefully classified as its production applying an ABC classification, and the periodicity of the units was defined by means of A, daily; B, two or three times a week; and C, one time a week. The biggest hospitals (almost 10% of all 78 units) of the city were characterized by heavy waste collection (about 1–4t) and 30–45 min handcraft removal. The concessionary had problems with covering the health units, because many of them caused claims from municipalities affected by the calendar and no service during the week. The solution reduced in 5% in the total distance traveled, but the number of health units went from 72 to 78, and the trips that were 2–3 per day performed in 10–12 h by the same staff (one driver and three laborers) went to two shifts of 8 h in three routes using two drivers and six laborers. The increase in cost compensated in the reality of the process, because the only driver and his laborers were not supporting such amount of work.

11.6 Waste Management DSS Infrastructure with Sectoring, Routing, and Scheduling

Decision support with IT infrastructure has been developed along the years to better capture the many different logistic concepts, as explained in the sections above.

Once waste collection is basically anchored in reverse-logistic transportation, the actors of this structure affect most part of any waste collection management system.

For any DSS in waste collection management (WCM), it is important to deal with the following aspects:

1. GIS-based system, with tools for network analysis and regional waste collection production analysis capabilities [70].
2. Data management of actual infrastructure available (vehicles, equipment, people) to perform coverage of the areas and services required (domiciliary, selective, health waste, sweeping, industrial-commercial, fair, etc.);
3. Data management and analytics for the regions of coverage and their routes to be performed per service and their schedule;
4. Data management for IoT and telemetry monitoring [4].
5. Data management to follow the executions per service and compare planned vs. executed services;
6. Analytics over the scenarios to hold field occurrences, reformatting regions and recalculating routes;
7. Data management to follow the resources expenses.

As previously mentioned, a number of applications were developed with much success in this field for different contexts of waste collection. In Table 11.3, we consider a summary of them, where the items' features can be described as follow:

Application: Refers to the type of service the software can be applied;
Resources: Indicates if the software can control or consider the use of resources (people and vehicles) to schedule in conjunction to be assigned to the routes;
Plan: Refers to the type of plans designed by the software to all applications attended;
IoT Support: Refers to the internet of things (IoT) service available by the software to monitor vehicles, bins, and/or crews in the field by using telemetry (equipment that permits dynamic measures of location, traveled distance, time, acceleration, velocity, breaks actions, hydraulic pressure, vehicle interventions, stops, etc.), and following the execution of the service (routes) according to the street-by-street sequence defined by the optimized plans;
Costs: Refers to the software tools available to follow the costs/expenses with the crews and vehicles during daily execution;
Rebuild: Refers to the software tools available to recalculate the actual routes because of network changes (street orientation, turn prohibitions, street segment, or corner temporarily prohibited), coverage changes (coverage or limits of sectors and circuits).

Negreiros et al. [93] presented the detailed architecture of Graphvs' WCMS OPILM-SisRot® LIX. In Fig. 11.15, we have a new visualization of the architecture, where it is included in two levels the major systems used, both installed in a VPS (web - cloud) that can be adequately dimension to attend a city or a concessionary (operating in different cities).

Table 11.3 Main resources present in WCM systems investigated

Software	WCM Characteristics					
	Application	Resources	Plan	IoT	Costs	Rebuild
ESRI™	Domic.	No Control	Distr. and Routes	Has tools	No	Yes
RouteSmart™	Domic. Selective Sweeping	No Control	Sectors-Routes	Yes	No	Yes
WasteRoute™	Commercial	No Control	Districts and Routes	Yes	No	Yes
FleetRoute™	Domic. Selective Commercial	No Control	Sectors-Routes	Yes	No	Yes
SCOLDSS	Domic. Selective Commercial	No Control	Sm. circuits & Routes	No	No	Yes
SisRot® LIX	Domic. Selective Sweeping Commercial Health	Control	Sectors-Routes	Yes	Yes	Yes
Opitrak™	Domic. Selective	No Control	Districts & Routes	Yes	No	Yes
Rapidis™	Domic.	No Control	Districts & Routes	Yes	No	Yes
RouteSW	Domic. Selective	No Control	Routes	No	No	Yes
Caliper TCAD™	Domic. Selective	No Control	Districts & Routes	No	No	Yes
SIRIUS	Domic. Selective Industrial Commercial Health	Control	Routes	Yes	Yes	Yes

In the main first level we have the software planner, SisRot®LIX. The software contains tools to postgres-SQL data management system, the graph-based modeling system, the GIS-integrated and visualizer, frequency assignment data management tools, the load/perimeter constrained clustering tools to define macro-regions, zones (areas with sectors with same frequency), sectors (area where a teams of crew perform their workload), and circuits (area where a vehicle perform one trip), the constrained general routes tools (capacitated by maximum waste offer and/or maximum workload) and movement prohibition reduction tools, for domiciliary or street sweeping, the 3D special points (customers, containers, fairs, etc.) identifier, the standard visualizer of covering and routes solutions on cartographic and satellite digital maps, the tools to manage the assignment of teams (drivers and collectors,

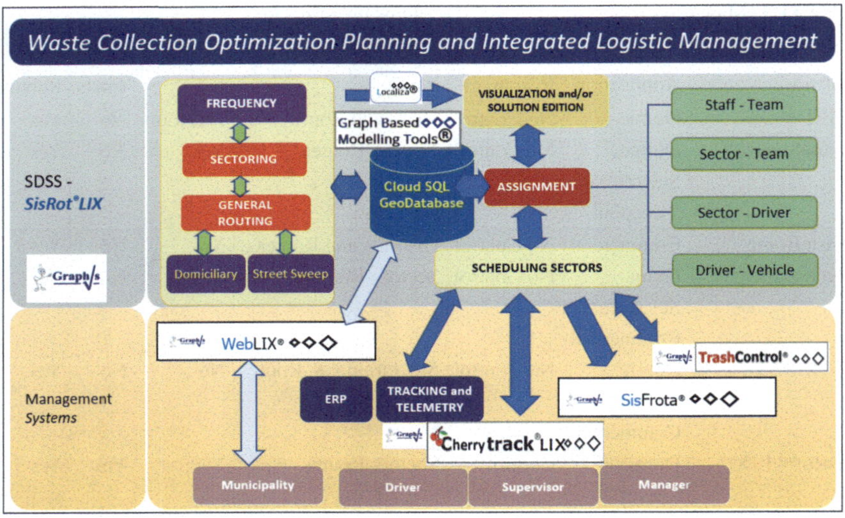

Fig. 11.15 WCMS OPILM-SisRot® LIX architecture designed by GRAPHVS Ltda

crew-vehicle, crew-sector), and finally the sector scheduler to send the results to the cloud.

In the second level, there are systems to control the daily production and expenses with people (ERP, TrashControl, and WebLix), control the expenses of the fleet (ERP, SisFrota), and control and follow the execution of the plans (CherryTrack Lix) with the mobile apps Driver, Supervisor, Manager, and the one to municipally follow the proximity of the vehicle to its home address or days and hours of service in his region. Tools for monitoring and perform telemetry are accepted here. Other ERP tools can be also connected if it is possible to transfer data between storage units by ETL.

11.7 Conclusion

This chapter studied waste collection reverse logistic systems actually practiced worldwide, and their relation to diverse OR approaches in districting and sectoring, node, arc, and both routing. The diverse solution methodologies showed us that each city has its own way to do the service because of no social policies and municipality claims.

Main reasons to evaluate the efficiency of each system, and perform savings evaluation, are concentrated in tools to best fit to the system in use. In all cases, there is no study to evaluate the best approach to be used for the context and organize the whole process accordingly. This work opens an opportunity to conceive adequate process and establish the choices of good managing and planning with minimal costs.

It is easy to see that the adoption of the bin and container systems is preferable than the street by street or any other, and it is a point that the use of classic VRP tools is more evident. In any case, it is gain force along the years the arc routing approaches especially the sector with MCARP with intermediate facilities as the routing step.

The sectoring and routing problems are rich in details and data to better approximate the solutions to the reality. This, for sure, will be the source of the success of any study application.

The study went to domiciliary, selective, commercial, and industrial waste collection and explored this diversity of services in many countries. For the end, we gave a vision of a waste collection management system that used such mathematical knowledge-based system and expert evolved by the heuristics, metaheuristics, and math-heuristics that can be used for this propose.

It is still open opportunities to better understand and propose adequate waste management systems according to the cities' needs and policies. For the end, worries on sustainability are guaranteed to be solved, with high impact in cost reductions and in the environment.

Acknowledgments We would like to thank CNPq (Grant 407652/2022-1), and GRAPHVS Ltd, which made this research possible. Special thanks to Augusto Palhano and Gerson Alves de Castro for their contribution in preparing XNes® and SisRot® Lix for evaluating different network instances here discussed.

Competing Interests This study was funded by [CNPq].

The authors have no conflicts of interest to declare that are relevant to the content of this chapter.

References

1. I.S. Abdelli, F. Abdelmalek, A. Djelloul, K. Mesghouni, A. Addou, GIS-based approach for optimized collection of household waste in Mostaganem city (Western Algeria). Waste Manag. Res. **34**(5), 417–426 (2016). https://doi.org/10.1177/0734242X16633519
2. S.M. Ahmed, Using GIS in solid waste management planning: a case study for Aurangabad, India. Linköpings University, Master Thesis, 2006
3. D. Ahr, G. Reineld, The capacitated arc routing problem: combinatorial lower bounds, Arc Routing: problems, methods and applications, in *MOS-SIAM Series on Optimization*, ed. by A. Corberan, G. Laporte, Chapter 8 (2014), pp 159–181
4. T. Anagnostopoulos, K. Kolomvatsos, C. Anagnostopoulos, A. Zaslavsky, S. Hadjiefthymiades, Assessing dynamic models for high priority waste collection in smart cities. J. Syst. Softw. **110**, 178–192 (2015). https://doi.org/10.1016/j.jss.2015.08.049
5. R. Assaf, Y. Saleh, Vehicle-Routing optimization for municipal solid waste collection using genetic algorithm: the case of southern Nablus city. Civil Environ. Eng. Rep. **26**(3), 043–057 (2017). https://doi.org/10.1515/ceer-2017-0034
6. J. Bautista, E. Fernández, J. Pereira Solving an urban waste collection problem using ants heuristics. Comput. Oper. Res. **35**, 3020–3033 (2008)

7. E. Beltrami, L. Bodin, Networks and vehicle routing for municipal waste collection. Networks **4**(1), 65–94 (1974)
8. J.M. Belenguer, E. Benavent, S. Irnich, The capacitated arc routing problem: exact algorithms. Arc Routing: problems, methods and applications, im *MOS-SIAM Series on Optimization*, ed. by A. Corberan, G. Laporte, Chapter 9 (2014), pp. 183–221
9. E. Benavent, D. Soler, The directed rural postman problem with turn penalties. Transp. Sci. **33**, 408–418 (1999)
10. J. Beliën, L. Boeck, J.V. Ackere, Municipal solid waste collection and management problems: a literature review. Transp. Sci. **48**(1), 78–102 (2014). https://doi.org/10.1287/trsc.1120.0448
11. R. Benitez-Bravo, R. Gomez-González, P. Rivas-Garcia, J.E. Botello-Alvarez, J.F. Huerta-Guevara, A.M. Garcia-Leon, J.F. Rueda-Avellaneda, Optimization of municipal solid waste collection routes in a Latin-American Context. J. Air Waste Manag. Assoc. **71**(11), 1415–1427 (2021) https://doi.org/10.1080/10962247.2021.1957040
12. E. Betanzo-Quezada, M.A. Torres-Gurrola, J.A. Romero-Navarrete, S.A. Obregón-Briosca, Evaluación de rutas de recolección de residuos sólidos urbanos con apoyo de dispositivos de rastreo satelital: análisis e implicaciones. Rev. Int. Contam. Ambie. **32**(3), 323–337 (2016) https://doi.org/10.20937/RICA.2016.32.03.07
13. F. Bertero, M. Cerdero, G. Dura'n, N.A. Fallace, Optimización de la zonificación de resíduos de um munícipio del Gran Buenos Aires. Ingeniería de Sistemas **35**, 49–80 (2021)
14. T. Bianchi-Aguiar, M.A. Carravilla, J.F. Oliveira, Municipal waste collection in Ponte de Lima, Portugal—a vehicle routing application. OR Insight **25**(4), 185–198 (2012). https://doi.org/10.1057/ori.2011.23
15. N. Bianchessi, A. Corberán, I. Plana, M. Reula, J-M. Sanchis, The profitable close-enough Arc Routing problem. Comput. Oper. Res. **140**, 105653 (2022)
16. L.D. Bodin, S. Kursh, A computer assisted-system for routing and scheduling of street sweepers. Oper. Res. **26**, 525–537 (1978)
17. L.D. Bodin, L. Levy, The arc-oriented location routing problem. INFOR **27**, 74–94 (1989)
18. L.D. Bodin, L. Levy, The arc partitioning problem. EJOR **53**, 393–401 (1991)
19. F. Bonomo, G. Durán, F. Larumbe, J. Marenco, A method for optimizing waste collection using mathematical programming: a Buenos Aires case study. Waste Manag. Res. **30**(3), 311–324, (2012). https://doi.org/10.1177/0734242X11402870
20. B. Boyaci, T.H. Huong, A. Lechford, Fast upper and lower bounds for a large-scale real-world arc routing problem. Networks **81**, 107–124 (2023)
21. C. Blazquez, G. Paredes, Network design of a household waste collection system: a case study of the commune of Renca in Santiago, Chile. Waste Manag. **116**, 179–189. (2020). https://doi.org/10.1016/j.wasman.2020.07.027
22. G. Braier, G. Durán, J. Marenco, F. Wesner, An integer programming approach to a real-world recyclable waste collection problem in Argentina. Waste Manag. Res. **35**(5), 525–533 (2017). https://doi.org/10.1177/0734242X16688
23. O. Bräysy, E. Martínez, Y. Nagata, D. Soler, The mixed capacitated general routing problem. Expert Syst. Appl. **38**, 12954–12966 (2011)
24. A. Butsch, J. Kalcsics, G. Laporte. Districting for arc routing. INFORMS J. Comput. **26**(4), 809–824 (2014)
25. Caliper, TransCAD: transportation planning software, Brochure (2023). https://www.caliper.com/transcad/applicationmodules.htm. Accessed 27 Oct 2023
26. M. Caramia, D.M. Pinto, E. Pizzari, G. Stecca, Clustering and routing in waste management: a two-stage optimization Approach. EURO J. Transp. Logist. (2023). https://doi.org/10.1016/j.ejtl.2023.100114
27. V. Del Carmen-Niño, R. Herrera-Navarrete, A.L. Juárez-López, M.L. Sampedro-Rosas, M. Reyes-Umaña, Municipal solid waste collection: challenges, strategies and perspectives in the optimization of a municipal route in a Southern Mexican Town. Sustainability **15**, 1083. (2023). https://doi.org/10.3390/su15021083
28. A. Cavallin, D. Rossit, V. Herrán, D. Rossit, M. Frutos, Application of a methodology to design a municipal waste pre-collection network in real scenarios. Waste Manag. Res. **38**(1), 117–129. (2020). https://doi.org/10.1177/0734242X19894630

29. C. Chalkias, K. Lasaridi, A GIS based model for the optimisation of municipal solid waste collection: the case study of Nikea, Athens, Greece. WSEAS Trans. Environ. Dev. **5**(10), 640–650
30. C. Chalkias, K. Lasaridi, Benefits from GIS based modelling for municipal solid, in *Waste Management, Integrated Waste Management*, ed. by Sunil Kumar, vol. I (InTech, 2011). ISBN: 978-953-307-469-6
31. CIVIX L.L.C (2024). FleetRouteTM. https://www.fleetroute.com/k1/e.php. Accessed 9 Jan 2024
32. J. Clossey, G. Laporte, P. Soriano, Solving arc routing problems with turn penalties. J. Oper. Res. Soc. **52**, 433–439 (2001)
33. A. Corberán, R. Martí, A. Romero, Heuristics for the mixed rural postman problem. Comput. Oper. Res. **27**, 183–203 (2000)
34. A. Corberán, A. Lechford, J.-M. Sanchis, A cutting plane algorithm for the General Routing Problem. Math. Program. A **90**, 291–316 (2001)
35. A. Corberán, R. Marti, E. Martinez, D. Soler, The Rural Postman Problem on mixed graphs with turn penalties. Comput. Oper. Res. **29**, 887–903 (2002)
36. A. Corberán, A. Romero, J.M. Sanchis, The mixed general routing polyhedron. Math. Program. **96**, 103–137 (2003)
37. A. Corberán, L. Plana, J.M. Sanchis, The chinese postman problem on directed, mixed and windy graphvs. Arc Routing: Problems, methods and applications, in *MOS-SIAM Series on Optimization*, ed. by A. Corberan, G. Laporte, Chapter 4 (2014), pp. 65–83
38. A. Corberán, I. Plana, J.M. Sanchis, The rural postman problem on directed, mixed, and windy graphvs. Arc Routing: Problems, methods and applications, in *MOS-SIAM Series on Optimization*, ed. by A. Corberan, G. Laporte, Chapter 6 (2014), pp. 101–127
39. A. Corberán, I. Plana, M. Reula, J.-M. Sanchis, A matheuristic for the distance-constrained close-enough Arc Routing Problem. TOP **27**, 312–326 (2019)
40. M.J Cortinhal, M.C. Mourão, A.C. Nunes, Local search heuristics for sectoring routing in a household waste collection. Eur. J. Oper. Res. 1–12, (2016). https://doi.org/10.1016/j.ejor.2016.04.013
41. A. Del Pia, C. Filipe, A variable neighborhood descent algorithm for a real waste collection problem with mobile depots. ITOR **13**, 125–141 (2006)
42. H.S. Düzgün, S. Uşkay, A. Aksoy, Parallel hybrid genetic algorithm and GIS-based optimization for municipal solid waste collection routing. J. Comput. Civil Eng. **30**(3), 04015037 (2016). https://doi.org/10.1061/(ASCE)CP.1943-5487.0000502
43. H.A. Eiselt, M. Gendreau, G. Laporte, Arc routing problems (Part I): the chinese postman problem. OR **43**, 231–242 (1995)
44. H.A. Eiselt, M. Gendreau, G. Laporte, Arc routing problems (Part II): the rural postman problem. OR **43**, 399–414 (1995)
45. E. Fadda, L. Gobbato, G. Perboli, M. Rosano, R. Tadei, Waste collection in urban areas: a case study. Interfaces **48**(4), 307–322 (2018)
46. M. Faccio, A. Alessandro Persona, G. Zanin, Waste collection multi objective model with real time traceability data. Waste Manag. **31**, 2391–2405 (2011)
47. A. Gallardo, M. Carlos, M. Peris, F.J. Colomer, Methodology to design a municipal solid waste generation and composition map: a case study. Waste Manag. **34**, 1920–1931 (2014). https://doi.org/10.1016/j.wasman.2014.05.014
48. G. Ghiani, F. Guerriero, G. Improta, R. Musmanno, Waste collection in Southern Italy: solution of a real-life arc routing problem. Intl. Trans. Opt. Res. **12**, 135–144 (2005)
49. G. Ghiani, D. Laganà, E. Manni, R. Musmanno, D. Vigo, Operations research in solid waste management: a survey of strategic and tactical issues. Comput. Oper. Res. **44**, 22–32. https://doi.org/10.1016/j.cor.2013.10.006
50. G. Ghiani, M.C. Mourão, L. Pinto, D. Vigo, Routing in waste collection applications. Arc Routing: problems, methods and applications, in *MOS-SIAM Series on Optimization*, ed. by A. Corberán, G. Laporte, Chapter 15 (2014), pp. 351–370

51. M. Gil, G.M. Avila, Comparative study of means of transport used in selective collection. ANAP-Cidades Verdes **5**(11), 61–74. ISSN: 2317.8604
52. A. Gilardino, J. Rojas, H. Mattos, G. Larrea, I. Vázquez, Combining operational research and life cycle assessment to optimize municipal solid waste collection in a district in Lima (Peru). J. Cleaner Prod. **156**, 589–603 (2017). https://doi.org/10.1016/j.jclepro.2017.04.005
53. B. Golden, R. Wong, Capacitated arc routing problem. Networks **11**, 305–315 (1981)
54. M. Gondran, M. Minoux, *Graphs and Algorithms*. Wiley Series in Discrete Mathematics and Optimization, 1st edn. (1984)
55. M. Grotschel, Z. Win, On the windy postman polyhedron, Techinical Report 75, University of Augsburg, Augsburg, Germany (1988).
56. S.A. Greco, D.G. Rossit, M. Frutos, A. Cavallin, Optimization of waste collection through the sequencing of micro-routes and transfer station convenience analysis: an Argentinian case study. Waste Manag. Res. (ISWA), **41**(7). https://doi.org/10.1177/0734242X221139123
57. M.G. Guan, Graphic programming using odd and even points. Chinese Math. **1**, 237–277 (1962)
58. M.G. Guan, On the windy postman problem. Discrete Appl. Math. **9**, 41–46 (1984)
59. M.-H. Há, N. Bostel, A. Langevin, J.-M. Rousseau, Solving the close enough arc routing problem. Networks **63**, 107–118 (2014)
60. H. Han, E. Ponce-Cueto, Waste collection vehicle routing problem: literature review. Traffic Transp. **27**(4), 345–358 (2015)
61. S. Hanafi, A. Fréville, P. Vaca, Municipal solid waste collection: an effective data structure for solving the sectorization problem with local search methods. INFOR **37**, 236–254 (1999)
62. K. Hauge, J. Larsen, R.M. Lusby, E. Krapper, A hybrid column generation approach for an industrial waste collection routing problem. Comput. Ind. Eng. **71**, 10–20 (2014)
63. C. Hess, A. Dragomir, K. Doerner, D. Vigo, Waste collection routing: a survey on problems and methods. Central Eur. J. Oper. Res. 1–36. https://doi.org/10.1007/s10100-023-00892-y
64. C. Hierholzer, Über die Möglichkeit, einen Linienzug ohne Wiederholung und ohne Unterbrechnung zu umfahren. Mathematische Annalen **6**, 30 (1873)
65. J. Janela, C. Mourão, L.S. Pinto, Arc routing with trip-balancing and attractiveness measures—a waste collection case study. Comput. Oper. Res. **147**, 105934 (2022)
66. J.P. Jacinto, R. Alvarenga-Rosa, R.S. Banos, Heurística para soluço do problema da coleta de resíduos sólidos domiciliares (RSD) com base no problema do carteiro chinós capacitado com múltiplas viagens (PCCC-MV). Transportes **22**(1) (2014). ISSN: 2237–1346
67. A. Kallel, M.M. Serbaji, M. Zairi, Using GIS-based tools for the optimization of solid waste collection and transport: case study of Sfax City, Tunisia. J. Eng. 4596849. https://doi.org/10.1155/2016/4596849
68. N.V. Karadimas, V.G. Loumos, GIS-based modelling for the estimation of municipal solid waste generation and collection. Waste Manag. Res. **26**, 337–346. https://doi.org/10.1177/0734242X07081484
69. C.H. Kappauf, Koehler, The mixed postman problem. Discrete Appl. Math. **1**, 89–103 (1979)
70. B-I. Kim, S. Kim, S. Sahoo, Waste collection vehicle routing problem with time windows. Comput. Op Res. **33**(12), 3624–3642 (2006). https://doi.org/10.1016/j.cor.2005.02.045
71. J.R. Kinobe, T. Bosona, G. Gebresenbet, C.B. Niwagaba, B. Vinnerås, Optimization of waste collection and disposal in Kampala city. Habitat Int. **49**, 126–137 (2015). https://doi.org/10.1016/j.habitatint.2015.05.025
72. P. Lacomme, C. Prins, W. Rendane-Chérif, Competitive memetic algorithms for arc routing problems. Ann. Oper. Res. **131**(4), 156–185 (2006)
73. J.-Q. Li, D. Borenstein, P.B. Mirchandani, Truck scheduling for solid waste collection in the City of Porto Alegre, Brazil. Omega **36**, 1133–1149 (2008). https://doi.org/10.1016/j.omega.2006.04.007
74. J.-Q. Li, D. Borenstein, P.B. Mirchandani, Truck schedule recovery for solid waste collection in Porto Alegre, Brazil. ITOR **15**, 565–582 (2008)
75. R.S. Lima, J.P. Lima, T.V. Vito-Silva, Arc routing using geographic information system for transportation: an application in urban solid waste system. J. Transp. Literature **6**(2), 180–196 (2012)

76. E. Lucas, Récréacions mathématiques IV, Paris, in *Graph Theory with Applications*, ed. by J.A. Bondy, U.S. Murty, 5th print (North Holland, Amsterdam, 1982)
77. L. Madahvi, S. Mansur, M.S. Sajadieh, Sustainable multi-trip periodic redesign-routing model for municipal solid waste collection network: the case study of Tehran. Environ. Sci. Pollut. Res. **29**, 35944–35963 (2022). https://doi.org/10.1007/s11356-021-18347-9
78. A. Mahéo, D.G. Rossit, P. Kilby, Solving the integrated bin allocation and collection routing problem for municipal solid waste: a Benders decomposition approach. Ann. Oper. Res. **322**, 441–465. https://doi.org/10.1007/s10479-022-04918-7
79. H. Mation, Caminhos eulerianos de mínimo custo em grafos eulerianos. Anais do XV SBPO e I CLAIO 483–492 (1982)
80. H. Mation, Um modelo de distribuiço aplicado ao problema de transporte de resíduos gerados em grandes cidades, MSc Dissertation, COPPE-Eng Produço/UFRJ (1984).
81. J.A.S. Martinez, A. Mendoza, M.R. Vazquez, Collection of solid waste in municipal areas: urban logistics. Sustainability **11**, 5442 (2019). https://doi.org/10.3390/su11195442
82. R. McBride, Controlling left and U-turns in the routing of refuse collection, Comput. Oper. Res. **9**(2), 145–152 (1982)
83. S. Moazzeni, M. Tavana, S.M. Darmian, A dynamic location-arc routing optimization model for electric waste collection vehicles. J. Cleaner Prod. **364**(1), 132571 (2022). https://doi.org/10.1016/j.jclepro.2022.132571
84. D.C. Mossige, S.L. Vigerust, An Evolutionary Algorithm for Waste Collection in Complex City Environments. Industrial Economics and Technology Management, PhD Thesis. Norwegian University of Science and Technology (2018)
85. M.C. Mourão, A.C. Nunes, C. Prins, Heuristic methods for the sectoring arc routing problem. EJOR **196**, 856–868 (2009)
86. M.J. Negreiros, Contribuições para otimização em grafos e o problema de percurso de veículos: Sistema SisGRAFO. PhD Thesis, COPPE-UFRJ, Rio de Janeiro/Brazil (1996)
87. M.J. Negreiros, GTSP based methods to the mixed rural postman problem. J. Heuristics (2025, submitted)
88. M.J. Negreiros, N. Maculan, D. Claiman, Optimizing routes in household refuse collection (a field application using SisRot), COPPE-UFRJ Technical Report ES-255/92
89. M. Negreiros, W. Coelho, Jr., A.W. Palhano, E. Coutinho, G.A. Castro, F.J. Negreiros, G. Barcellos, B. Rezende, L. Pereira, O problema do carteiro chinês, algoritmos exatos e um ambiente mvipara análise de suas instâncias: Sistema XNÊS. Pesquisa Operacional **29**(2), 323–363 (2009)
90. M.J. Negreiros, Palhano: line-graph transformation to the Wuler tour with movement prohibition problem. Annals of IFORS, Melbourne/AU (2011)
91. M. Negreiros, N. Maculan, P. Batista, A. Rodrigues, A. Palhano, Capacitated clustering problems applied to the layout of it-teams in software factories. Ann. Oper. Res. **316**, 1157–1185 (2020). https://doi.org/10.1007/s10479-020-03785-4
92. M.J. Negreiros, A.W.C. Palhano, E.C. Reis, Sector arc routing-based spatial decision support system for waste collection in Brazil. Waste Manag. Res. **1**, 0734242X2211043
93. M.J. Negreiros, A.W.C. Palhano, E.C. Reis, Optimized Planning and Management of domiciliary and selective solid waste: Results of application in brazilian cities(SisRot LIX). Solid Waste and Landfills Management Recent Advances, Suhaiza Zailani and Suriyanarayanan Sarvajakesavalu Ed, IntechOpen, Chapter 2, pp 25–56, DOI:10.5772/intechopen.100756.
94. M. Negreiros, N. Maculan, P. Michelon, R.V. Pinto, New generalization to the capacitated centred clustering problem. SBPO/São José dos Campos-SP, Anais do LV SBPO (2023)
95. M.J. Negreiros, A.W. Palhano, N. Maculan, C. Barbieri, The twin-rural postman problem. EURO 2024, Annals (2024)
96. M.J. Negreiros, N. Maculan, A. Palhano, P. Batista, The heterogeneous arc sector routing problem. Comput. Eng. (2025, submitted)
97. T. Nuortio, J. Kytöjoki, H. Niska, O. Bräysy, Improved route planning and scheduling of waste collection and transport. Expert Syst. Appl. **30**, 223–232 (2006)
98. J.K. Odum, A. Owuso-Addo, N. Kyere-Sacrifice, A.-A. Kwarteng, A construction of routes for vehicles using the transportation system of solid waste in the Penkwase. Int. J. Multidiscip. Stud. Innov. Res. **8**, 535–552 (2022). https://doi.org/10.53075/Ijmsirq065465721

99. T.C. Ogwueleka, Route optimization for solid waste collection: Onitsha (Nigeria) case study. J. Appl. Sci. Environ. Manag. **13**(2), 37–40 (2009)
100. C.S. Orloff, A fundamental problem in vehicle routing. Networks **4**, 35–64 (1974)
101. Optrak Software: OptrakTM Vehicle Routing Software. https://optrak.com/commercialwaste/. Accessed 10 Jan 2024
102. Optit: OptiRouteTM Waste. https://www.optit.net/. Accessed 10 Jan 2024
103. C. Prins, The capacitated arc routing problem. Arc Routing: Problems, methods and applications, in *MOS-SIAM Series on Optimization*, ed. by A. Corberán, G. Laporte, Chapter 7 (2014), pp 131–127
104. S. Sahoo, S. Kim, B.-I. Kim, B. Kraas, A. Popov, Jr., Routing optimization for waste management. Interfaces **35**(1), 24–36 (2005)
105. L. Santos, J. Coutinho-Rodrigues, Implementação em SIG de uma Heurística para o Estudo da Recolha da Resíduos Sólidos Urbanos, Relatórios INESC (2003)
106. J. Sniezek, L. Bodin, Using mixed integer programming for solving the capacitated arc routing problem with vehicle/site dependencies with an application to the routing of residential sanitation collection vehicles. Ann. Oper. Res. **144**, 33–58 (2006)
107. J. Sniezek, J.; L. odin, L. Levi, M. Ball, The capacitated vehicle routing problem with vehicle/site dependencies: the Philadelphia experience, in *The Vehicle Routing Problem*, ed. by P. Toth, D. Vigo. SIAM Monographs on Discrete Mathematics and Its Applications, Chapter 9 (2002), pp 287–308
108. Rapidis, Route Planning for Waste Collection (2023). https://www.rapidis.com/route-planning-waste-collection/?lang=en. Accessed 8 Jan 2024
109. M. Reula, R. Martí, Heuristics for the profitable close-enough arc routing problem. Expert Syst. Appl. **230**, 120513 (2023)
110. A.M. Rodrigues, J.S. Ferreira, Waste collection routing—limited multiple landfills and heterogeneous fleet. Networks **65**(2), 155–165
111. RouteSmart Technologies (2024) RouteSmartTM. https://www.routesmart.com/. Accessed 8 Jan 2024
112. S. Roy, J.-M. Rousseau, The capacitated Canadian postman problem, in *INFOR*, vol. 27 (1989), pp 58–73
113. L. Santos, J. Coutinho-Rodrigues, J. Current, Implementing a multi-vehicle multi-route spatial decision support system for efficient trash collection in Portugal. Transp. Res. A **42**, 922–934 (2008)
114. E.O. Simonetto, D. Borenstein, A decision support system for the operational planning of solid waste collection. Waste Manag. **27**, 1286–1297 (2007)
115. SIRIUS-es, Solids Waste Management (2024). https://www.siriusgemstone-es.com/services-2/environmentalsolution/solid-waste-management/. Accessed 26 Apr 2024
116. L.H. Son, A. Louati, Modeling municipal solid waste collection: a generalized vehicle routing model with multiple transfer stations, gather sites and inhomogeneous vehicles in time windows. Waste Manag. **52**, 34–49 (2016). https://doi.org/10.1016/j.wasman.2016.03.041
117. A. Sulemana, A.E. Donkor, E. Forkuo, S. Odoru-Kwarteng, Optimal routing of solid waste collection trucks: a review of methods. J. Eng. **2018**, 4586376. https://doi.org/10.1155/2018/4586376
118. A. Sulemana, E.A. Donkor, E.K. Forkuo, J. Asantewaa, I.N.A. Ankrah, A.M.O. Musah, Optimized routing of trucks for institutional solid waste collection in Kumasi, Ghana. Detritus **9**, 50–58. https://doi.org/10.31025/2611-4135/2020.13896
119. G. Tavares, Z. Zsigraiova, V. Semiao, M.G. Carvalho, Optimisation of MSW collection routes for minimum fuel consumption using 3D GIS modelling. Waste Manag. **29**(3), 1176–1185 (2009)
120. G. Ulusoy, The fleet size and mix problem for capacitated arc routing. EJOR **22**, 329–337 (1985). https://doi.org/10.1016/0377-2217(85)90252-8
121. T.P.B. Vecchi, Otimização de rotas dos caminhões de coleta de resíduos sólidos urbanos, PhD Thesis, UE Maringá – Brazil (2016)

122. T.P.B. Vecchi, D.F. Surco, A.C. Constantino, M.T.A. Steiner, L.M.M. Jorge, M.A.S.S. Ravagnani, P.R. Paraíso, A sequential approach for the optimization of truck routes for solid waste collection. Process Safety Environ. Protect. **102**, 238–250. https://doi.org/10.1016/j.psep.2016.03.014
123. T. Vidal, T.G. Crainic, M. Gendreau, N. Lahrichi, W. Rei, Oper. Res **60**(3) (2012). https://doi.org/10.1287/opre.1120.1048
124. H.L. Vu, K.T.W. Ng, D. Bolingbroke, Parameter interrelationships in a dual phase GIS-based municipal solid waste collection model. Waste Manag. **78**, 258–270 (2018)
125. A. Viktorin, D. Duáan Hrabec, V. Nevrlü, R. Šomplák, R. Šenkeřík, Hierarchical clustering-based algorithms for optimal waste collection point locations in large-scale problems: a framework development and case study. Comput. Ind. Eng. https://doi.org/10.1016/j.cie.2023.109142
126. E. Willemse, J. Joubert, Benchmark sets for variants of the capacitated arc routing problem under time restrictions with intermediate facilities. Data Brief **v8**, 972–977 (2016) https://doi.org/10.1016/j.orl.2016.06.001
127. E.J. Willemse, J.W. Joubert, Efficient local search strategies for the mixed capacitated arc routing problems under time restrictions with intermediate facilities. Comput. Oper. Res. **105**, 203–225 (2019)

The manufacturer's authorised representative in the EU is Springer Nature Customer Service Centre GmbH, Europaplatz 3, 69115 Heidelberg, Germany. If you have any concerns regarding our products, please contact ProductSafety@springernature.com

Printed and bound by CPI Group (UK) Ltd, Croydon, CR0 4YY
30/09/2026
02225584-0008